Advances in Experimental Medicine and Biology

Clinical and Experimental Biomedicine

Volume 1289

More information about this subseries at http://www.springer.com/series/16003

Mieczyslaw Pokorski
Editor

Medical and Biomedical Updates

Editor
Mieczyslaw Pokorski
Opole Medical School
Opole, Poland

The Jan Dlugosz University in Czestochowa
Czestochowa, Poland

ISSN 0065-2598 ISSN 2214-8019 (electronic)
Advances in Experimental Medicine and Biology
ISSN 2523-3769 ISSN 2523-3777 (electronic)
Clinical and Experimental Biomedicine
ISBN 978-3-030-67218-8 ISBN 978-3-030-67216-4 (eBook)
https://doi.org/10.1007/978-3-030-67216-4

This Springer imprint is published by the registered company Springer Nature Switzerland AG
The registered company address is: Gewerbestrasse 11, 6330 Cham, Switzerland

Contents

Adv Exp Med Biol - Clinical and Experimental Biomedicine (2021) 11: 1–25
https://doi.org/10.1007/5584_2020_561

Published online: 8 August 2020

Role of IP3 Receptors in Shaping the Carotid Chemoreceptor Response to Hypoxia But Not to Hypercapnia in the Rat Carotid Body: An Evidence Review

Anil Mokashi, Arijit Roy, Santhosh M. Baby, Eileen M. Mulligan, Sukhamay Lahiri, Camillo Di Giulio, and Mieczyslaw Pokorski

Abstract

This article addresses the disparity in the transduction pathways for hypoxic and hypercapnic stimuli in carotid body glomus cells. We investigated and reviewed the experimental evidence showing that the response to hypoxia, but not to hypercapnia, is mediated by 1,4,5-inositol triphosphate receptors (IP_3R/s) regulating the intracellular calcium content $[Ca^{2+}]_c$ in glomus cells. The rationale was based on the past observations that inhibition of oxidative phosphorylation leads to the explicit inhibition of the hypoxic chemoreflex. $[Ca^{2+}]_c$ changes were measured using cellular Ca^{2+}-sensitive fluorescent probes, and carotid sinus nerve (CSN) sensory discharge was recorded with bipolar electrodes in in vitro perfused-superfused rat carotid body preparations. The cell-permeant, 2-amino-ethoxy-diphenyl-borate (2-APB; 100 μM) and curcumin (50 μM) were used as the inhibitors of IP_3R/s. These agents suppressed the $[Ca^{2+}]_c$, and CSN discharge increases in hypoxia but not in hypercapnia, leading to the conclusion that only the hypoxic effects were mediated via modulation of IP_3R/s. The ATP-induced Ca^{2+} release from intracellular stores in a Ca^{2+}-free medium was blocked with 2-APB, supporting this conclusion.

Keywords

Carotid body · Carotid chemoreceptor responses · Carotid sinus nerve · Chemosensory discharge · Glomus cells · Hypercapnia · Hypoxia · IP_3 receptor · Phosphoinositides

A. Mokashi and S. Lahiri
Department of Physiology, University of Pennsylvania Medical Center, Philadelphia, PA, USA

A. Roy
Department of Physiology and Pharmacology, University of Calgary, Calgary, Canada

S. M. Baby
Department of Drug Discovery, Galleon Pharmaceuticals, Inc., Horsham, PA, USA

E. M. Mulligan
Pennsylvania State University, Altoona, PA, USA

C. Di Giulio
Department of Neurosciences Imaging and Clinical Sciences, University of Chieti-Pescara, Chieti, Italy

M. Pokorski (✉)
Institute of Health Sciences, University of Opole, Opole, Poland
e-mail: m_pokorski@hotmail.com

1 Introduction

Hypoxia increases carotid sinus nerve (CSN) sensory discharge emanating from the carotid body (CB). Based on the morphological and electrophysiological studies, CB glomus cells, in contact with sinus nerve endings, are considered the primary oxygen sensor (Lahiri et al. 2001; Gonzalez et al. 1995). Oxygen sensing involves both mitochondria (Wyatt and Buckler 2004; Mulligan et al. 1981) and cell membrane (Buckler and Vaughn-Jones 1998; Lopez-Barneo et al. 1988). During hypoxia, mitochondrial membrane depolarization has been observed in glomus cells (Roy et al. 2002). Duchen and Biscoe (1992) have suggested that the release of Ca^{2+} from intracellular stores, notably related to the endoplasmic reticular (ER) and mitochondria (MT), is indispensable for enhancing CSN discharge in response to hypoxia. Likewise, Lahiri et al. (1996) have reported the essential role of intracellular Ca^{2+} release for CSN stimulation during hypoxia, using thapsigargin, a blocker of Ca^{2+}-ATPase. In contradistinction, Lopez-Barneo et al. (1988) have proposed that cell membrane depolarization in hypoxia takes place in glomus cells due to the inhibition of oxygen-sensitive outward K^+ channels, followed by the influx of extracellular Ca^{2+} $[Ca^{2+}]_e$ causing, in turn, the release of neurotransmitters and enhanced CSN discharge. The plausible role of Ca^{2+} influx into the glomus cell in response to hypoxia has been pointed out in some other studies (Wyatt and Buckler 2004; Vicario et al. 2000), but there is increasing evidence that intracellular calcium content $[Ca^{2+}]_c$ associated with the ER-MT complex plays a key role in the process (Parekh 2003; Rutter and Rizzuto 2000; Rizzuto et al. 1999). Nonetheless, it is conceivable that activation of cell membrane-associated receptors and its effect on ER-MT-related $[Ca^{2+}]_c$ release, as well as additional Ca^{2+} influx via membrane ion/receptor channels, cooperate in hypoxia-induced activation of glomus cells.

Transmembrane signaling, leading to cellular excitation, usually involves the transmission of an external stimulus across the specific receptor proteins having innate enzyme activity, such as tyrosine kinase that transfers a phosphate group from ATP to cellular proteins (Hunter 2000). Protein phosphorylation, in general, also regulates the activation of membrane-associated ligand (receptor)-gated ion channels, including Ca^{2+} ion channels (Swope et al. 1999; Hain et al. 1995). Aside from protein phosphorylation, membrane trafficking includes receptors, protein docking-undocking, and the fusion machinery that shuttles between organelles in the cytoplasm and the cell membrane (Roth 2004). One of the major signaling pathways is the phosphatidylinositol (PI) cycle that consists of a lipid-associated conversion of PI to phosphatidylinositol phosphate (PIP), inositol 1,4-bisphosphate (PIP_2), and 1,4,5-inositol triphosphate (IP_3) and diacylglycerol (DAG) at the cell membrane site. Increased IP_3 formation and associated activation of IP_3 receptor sites (IP_3R/s) are crucial elements of the hydrolysis and phosphorylation processes in cells (Michikawa et al. 1996).

Both PI cycle and protein kinases are present in CB and their activation takes part in shaping the CB response to hypoxia, as reported in the cat and rat CB (Pokorski 2000; Faff et al. 1999; Pokorski and Faff 1999; Wang et al. 1999; Pokorski et al. 1996; Pokorski and Strosznajder 1993). Protein kinase C (PKC) increases CSN activity in the cat CB (Lahiri et al. 1990) and inhibits Ca^{2+}-dependent K^+ channels of glomus cells in the neonatal rat CB (Peers and Carpenter 1998). Thus, cellular stimulation with protein kinase activation via phosphorylation is bound to enhance IP_3 formation and IP_3R/s activation (Michikawa et al. 1996). IP_3R/s-mediated Ca^{2+} mobilization in hypoxia has been observed in neurons and PC12 cells (Kaplin et al. 1996). The observations of PKC-related activation of IP_3R/s in hypoxia have formed the groundwork for the role of IP_3R/s in oxygen-sensing mechanisms of excitable cells, notably of CB glomus cells.

Cells derived from a pheochromocytoma of the rat adrenal medulla (PC12 cell line) respond to hypoxic stimulus with the inhibition of membrane K^+ channels and secretion of neurotransmitters, notably dopamine, within less than 1 min of hypoxic exposure. These characteristic responses are akin to those of CB glomus cells (Kumar et al. 1998; Conforti and Millhorn 1997). PC12 cells also respond to hypoxic exposure (5% O_2) with enhanced phosphorylation and moderately increased expression of cAMP-response element-binding protein (CREB) in 20 min, reaching a peak level of expression after 6 h indicating hypoxia-induced increased C-fos and Jun-B gene expression (Beitner Johnson and Milhorn 1998). These delayed responses might not participate in the acute response to hypoxia. However, phosphorylation-associated activation of transmembrane proteins with receptor-gated (ligand-gated) ion channels at the cell membrane has a fast response (Swope et al. 1999).

Hypercapnia (10–20% inspired CO_2) also increases Ca^{2+} content in glomus cells, which is attributed to the cell membrane depolarization, without the participation of mitochondria (Buckler and Vaughn-Jones 1994; Peers 1990). Neither does hypercapnia appreciably change oxidative phosphorylation in the cerebral mitochondria of neonatal dogs (Yoshioko et al. 1995). Skeletal muscle mitochondrial ATP levels also are unchanged, separating the phosphorylation process from cellular hypercapnic effects (Thompson et al. 1992). Thus, it seems that increased phosphorylation activity is closely related to both short- and long-term hypoxia rather than to hypercapnia.

The effects of mitochondrial phosphorylation and phosphorylation at the cell membrane may not be mutually exclusive as mitochondrial inhibitors affect the glomus cell membrane potential (Roy et al. 2002; Buckler and Vaughn-Jones 1998). We have previously shown that inhibitors of oxidative phosphorylation block the increases in both $[Ca^{2+}]_c$ in glomus cells (Lahiri et al. 2003) and in in vivo CSN discharge (Mulligan et al. 1981) in response to hypoxia but not to hypercapnia. IP_3R/s induced Ca^{2+} mobilization in hypoxia and inhibition of such mobilization using heparin in PC12 cells points to a role of phosphorylation via IP_3R/s activation and inhibition in hypoxia (Kaplin et al. 1996). Those observations made us rationalize that 2-amino-ethoxy-diphenyl-borate (2-APB) and curcumin, cell membrane-permeant IP_3R/s blockers, as opposed to charged and membrane impermeant heparin, could yield a better insight into the role of membrane phosphorylation elicited by hypoxia. Both agents have been used as inhibitors in cerebellar microsomes (Dyer et al. 2002; Maruyama et al. 1997) and in Chinese hamster ovarian (CHO) cell line (Kukkonen et al. 2001). In the present article, we also describe a specific action of 2-APB on IP_3R/s in glomus cells using the ATP-induced activation of IP_3R/s in a Ca^{2+}-free solution. Additionally, we tested the caffeine-induced activation of ryanodine receptors (RyR/s) in the absence and presence of 2-APB: the receptors that are operational in the sarcoplasmic/endoplasmic reticulum membrane and are responsible for the release of Ca^{2+} from intracellular stores (Zaccheti et al. 1991).

2 Methods

2.1 Preparation and Protocol

To achieve the optimum hypoxic responses, we used an in vitro carotid body preparation that was superfused and perfused with solutions containing different PO_2 levels. These levels were akin to those used in other cellular studies and were in a range of 12–15 Torr in the superfusate (Roy et al. 2000; Duchen and Biscoe 1992) and 35–40 Torr PO_2 in the perfusate (Lahiri et al. 1996). For the hypercapnia-related investigations, PCO_2 of 55–60 Torr has been found adequate for both cellular and in vitro CB conditions (Roy et al. 2002; Buckler and Vaughn-Jones 1994).

The experimental protocol was designed to identify the role of the membrane-permeant IP_3R/s blockers 2-APB and curcumin. The following sets of experiments were done: (i) effects of membrane impermeant heparin on $[Ca^{2+}]_c$ and CSN discharge in hypoxia; (ii) effects of 2-APB and curcumin in the glomus cells during hypoxia and hypercapnia; (iii) effects of 2-APB and curcumin on CSN discharge during perfusion with hypoxic and hypercapnic solutions;

(iv) IP_3R/s activation with ATP and IP_3R/s inhibition with 2-APB; and (v) IP_3R/s inhibition with 2-APB on caffeine-induced activation of ryanodine receptors (RyR/s) and $[Ca^{2+}]_c$ increase.

2.2 Carotid Body Glomus Cells

The CB glomus cells from adult rats were separated enzymatically as described before (Roy et al. 2000). Briefly, the CBs were identified, surgically removed, and enzymatically dissociated in a calcium and magnesium-free buffer containing 0.1% collagenase and 0.1% hyaluronidase for 20 min with the continuous bubbling of a humidified gas mixture of 95% O_2 and 5% CO_2 at 37 °C. The digested CBs were then transferred to a sterilized test tube containing an F-12 growth medium (90 parts Ham F-12 and 10 parts fetal bovine serum) that was fortified with 100 units/mL of penicillin G and 100 μg/mL of streptomycin. The digested tissue was triturated with a fire-polished Pasteur pipette; the cells were plated on 18-mm coverslip and were kept undisturbed for 24 h in a Petri dish with air and 5% CO_2 circulating in a humidified incubator at 37 °C. The glomus cells were identified using the following criteria: (i) granular birefringent appearance; (ii) presence of large nuclei; and (iii) presence of monoamine-associated positive fluorescence using a sucrose-phosphate-glyoxylate test (De La Torre 1980).

2.3 IP_3R/s-Related Immunofluorescence in Glomus Cells in Hypoxia and Hypercapnia

Based on a report of Kaplin et al. (1996) indicating the role of IP_3R/s in PC12 cells during hypoxia, we used an IP_3R/s-targeted antibody to identify the presence of these receptors in glomus cells. The polyclonal antibody was produced using a short synthetic polypeptide that recognizes the C-terminal of the cytoplasmic domain of IP_3R/s (Santacruz Biotechnology; Dallas, TX). The antibody binding sites were made visible by binding a fluorescent marker to it. In this investigation, coverslips with 1–2-day-old glomus cells were used. The cells were superfused with hypoxic (PO_2 = 10 Torr) and hypercapnic (PCO_2 = 55 Torr, pH = 7.15) solutions in two sets of experiments. The cells were fixed in 4% paraformaldehyde in phosphate-buffered saline (PBS; pH = 7.4) for 10 min. After washing three times with PBS, cells were treated with 0.5% Triton solution for 5 min to permeabilize the membrane, followed by incubation with 10% fetal bovine serum in PBS for 1 h at room temperature to block non-specific binding sites. Then, cells were incubated with a saturated solution of the IP_3 receptor antibody for 2 h at room temperature, followed by incubation with a secondary rabbit IgG-fluorescein isothiocyanate-conjugated (FITC) antibody at 1:200 dilution. Following a thorough rinsing in PBS, coverslips were mounted in Mowiol (Calbiochem; San Diego, CA), and cell images were acquired by a charge-coupled digital CCD camera (Hamamatsu Photonics; Shizuoka, Japan). Cellular periphery was digitally marked and pixel intensities within the region were measured. The intensity of immunofluorescence in the control and stimulated cells was determined by subtracting the background intensity. Changes in immunofluorescence intensity due to hypoxia and hypercapnia were assessed in the control and stimulated cells in the absence and presence of 2-APB. Negative control was performed omitting the primary antibody and incubating the cells with fetal bovine serum alone, which failed to yield any immunofluorescence.

2.4 Ca^{2+} Measurements in Carotid Bodies Glomus Cells Exposed to Hypoxia and Hypercapnia

Glomus cells' responses were measured during superfusion with hypoxic and hypercapnic solutions. First, Fura-2 AM was used for the assessment of sensitive changes in $[Ca^{2+}]_c$. Indo-1 AM was also used to measure $[Ca^{2+}]_c$ changes in some of the experiments. Hypoxic and hypercapnic cellular responses were measured in the absence and presence of 2-APB and curcumin.

A normoxic normocapnic solution was used as a control in all of the measurements, according to the method described by Roy et al. (2000).

$[Ca^{2+}]_c$ Measurements with Fura-2 AM Cells on the coverslip were loaded with Fura-2 AM (5 μM) for 30–45 min at room temperature. The coverslip was placed in a closed chamber (Warner Instruments; Hamden, CT). A small volume of the chamber (70 μL) ensured a linear and rapid flow. The gravitational flow was adjusted to 2 mL/min. The chamber was mounted on a platform heated to 37 °C. Glomus cells were viewed with an inverted microscope. They were superfused with normoxic Tyrode's buffer for 15–20 min to remove the excess probe and to reduce background fluorescence. The cells were then excited at 340 nm and 380 nm, and the emission was set at 510 nm. Changes in a fluorescent ratio during hypoxic and hypercapnic superfusions in the absence and presence of 2-APB (100 μM) and curcumin (50 μM) were monitored continuously at a sampling rate of 30 frames/s. Data were computerized and were analyzed with Photon Technology Instrument software. Separate measurements of the fluorescent ratio were made in the control condition for up to 5 min to assess the photobleaching effect and to correct for the background fluorescence during the cellular response. Additional $[Ca^{2+}]_c$ investigations were done using the Indo-1 AM probe to obtain enough observations for statistical elaboration.

$[Ca^{2+}]_c$ Measurements with Indo-1 AM Cells on the coverslip were loaded with Indo-1 AM (5 μM) in HEPES buffer for 45 min at room temperature. The coverslip was attached to a flow-through chamber, and the cells were superfused with a gravity flow that was adjusted to 1.0–1.25 mL/min. Then, cells were excited with a xenon lamp light at 340 nm, and a fluorescence ratio (405/495 nm) was measured with PC and an analog recorder. Cellular responses were examined in 2–3 cells *per* field of view or a 3–5 cell cluster *per* field of view.

In case of heparin, due to its impermeant nature, glomus cells were loaded with Indo-1 AM and were pretreated for 30 min with heparin (1 mg/mL) in HEPES buffer before $[Ca^{2+}]_c$ measurements. Responses to normoxia (PO_2 = 125–130 Torr) and hypoxia (PO_2 = 12–15 Torr) at a constant pH (7.35–7.38; PCO_2 = 30–35 Torr) were measured before and after heparin pretreatment. Similar $[Ca^{2+}]_c$ measurements were made in the absence and presence of 2-APB during normoxia and hypoxia. Reversibility of hypoxic responses was tested with a return to superfusion with normoxic/normocapnic buffer, and it was used as a control measurement. The responses to an increased level of PCO_2 (from PCO_2 = 30–35 Torr, pH = 7.28–7.35 to PCO_2 = 50–55 Torr, pH = 7.15–7.18) at a constant PO_2 of 125–130 Torr were also taken. Changes in a fluorescence ratio were measured online with PC and a digitized recording device. $[Ca^{2+}]_c$ was calculated according to Grynkiewicz et al.'s (1985) formula: $[Ca^{2+}]_c = K_d\ (S_f/S_b)\ (R\text{-}R_{min})/(Rm_{ax}\text{-}R)$, where K_d is a dissociation constant, S_f is a concentration of free dye species at a given wavelength, S_b is for Ca^{2+}-bound dye, R is observed fluorescence ratio, and R_{min} and R_{max} are ratio values with 0 Ca^{2+} and 2.2 mM Ca^{2+}, respectively, with K_d of 250 nM for Indo-1 AM.

Changes in the fluorescence ratio were compared with a *t*-test, using the alpha significance level of 0.05.

2.5 Carotid Sinus Nerve (CSN) Activity in In Vitro Carotid Body (CB) During Hypoxia and Hypercapnia

The carotid artery bifurcation with the intact CB attached to the sinus nerve was removed and mounted in a perfusion chamber. The isolated CB was continuously perfused and superfused with a solution that had a similar composition of CO_2 and HCO^-_3 buffer to that used in cellular $[Ca^{2+}]_c$ studies at a constant pressure of 80 Torr and temperature of 37 °C. The effluent was continuously removed by suction. Sinus nerve afferent electrical activity was recorded with a bipolar electrode from the cut unsheathed end of the nerve, and data were fed into an online recorder and computer. Steady-state CSN responses were measured during normoxia ($PO_2 = 125$–130 Torr), hypoxia ($PO_2 = 35$–40 Torr), and hypercapnia ($PCO_2 = 50$–55 Torr) in the absence and presence of heparin (1 mg/mL), 2-APB (50–250 μM) and curcumin (25–50 μM). The perfusate PO_2 and PCO_2 were necessary to elicit suitable chemosensory responses (Roy et al. 2000; Lahiri et al. 1996).

All the experiments were performed in the presence of normal Ca^{2+} (2.2 mM) in the buffer. However, with ATP (100 μM)-induced IP_3R/s activation, some of the solutions were made with zero Ca^{2+} buffer that contained 1 mM ethylene-bis(oxyethylenenitrilo) tetraacetic acid (EGTA). These Ca^{2+}-free solutions were used to identify the role of intracellular Ca^{2+} stores in the release and global $[Ca^{2+}]_c$ increase. Calibrations for Indo-1 AM and Fura-2 AM have been described in full detail in a previous publication (Mokashi et al. 2003).

3 Results

3.1 Immunofluorescence of IP_3 Receptors

The presence of IP_3R/s was observed using the IP_3R-specific saturated solution of a polyclonal antibody, and it was visualized using a rabbit IgG conjugated with FITC antibody. A comparison of fluorescence between the control glomus cells and glomus cells that were superfused with a hypoxic solution (35–40 min) showed a significant increase in fluorescence due to hypoxia, pointing to an increased quantity of IP_3R/s formation. The cells that were treated with hypoxia in the presence of 2-APB (100 μM) showed a significant decrease in fluorescence, as the antagonist binding reduced the available IP_3R/s sites (Fig. 1a). When the cells were superfused with a hypercapnic solution, fluorescence failed to increase, remaining at a level akin to that during normocapnia and 2-APB did not change it. Thus, changes in IP_3R sites were inappreciable (Fig. 1b).

3.2 Glomus Cells' Calcium Content $[Ca^{2+}]_c$ and Carotid Sinus Nerve (CSN) Responses to Hypoxia Before and After Pretreatment with Heparin (Indo-1 AM Probe)

Change from normoxia ($PO_2 = 125$–130 Torr) to hypoxia ($PO_2 = 12$–15 Torr) in the superfusate increased $[Ca^{2+}]_c$ in glomus cells, from 94 ± 6 nM to 307 ± 45 nM ($n = 7$). Pretreatment

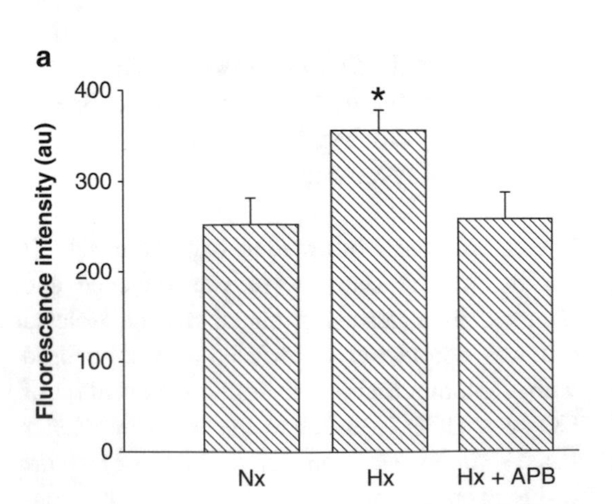

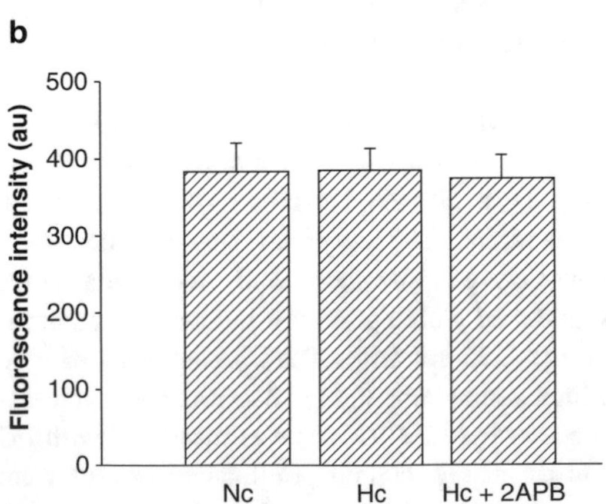

Fig. 1 Fluorescence intensity-related changes in glomus cells during hypoxia (**a**) and hypercapnia (**b**) in the presence and absence of 2-APB (100 μM) shown as means ±SE ($n = 7$). Significant increase (*$p < 0.05$) in the fluorescence intensity (in arbitrary units; au) was observed due to hypoxia alone, and it reverted to the normoxic level in the presence of 2-APB. In contradistinction, changes in the CO_2 level in the superfusate failed to appreciably influence the fluorescence intensity in glomus cells across the conditions investigated; *Nx* normoxia, *Hx* hypoxia, *2-APB,* 2-amino-ethoxy-diphenyl-borate; *Nc* normocapnia, *Hc* hypercapnia

of the same cells with heparin (1 mg/mL) for 30 min strongly diminished the hypoxia-related $[Ca^{2+}]_c$ increase, which was from 90 ± 9 nM to 130 ± 6 nM only (n = 8) (Fig. 2a).

In in vitro CB preparation, a change of perfusate from normoxia (PO_2 = 127 Torr) to hypoxia (PO_2 = 37 Torr) increased the chemosensory discharge in CSN from 55 imp/s to 355 imp/s. In the presence of heparin, the discharge increase was less, amounting to 285 imp/s at the same level of hypoxia (Fig. 2b). These results indicate that the cellular effects of heparin were prominent when compared to the effects emanating from the CB tissue source site. A large anionic charge associated with heparin might be responsible for the differential responses to hypoxia.

3.3 Effect of 2-APB on Hypoxic and Hypercapnic Responses (Fura-2 AM Probe)

During the superfusion of glomus cells with a hypoxic solution (PO_2 = 12–15 Torr) for 3 min, the fluorescence ratio (340/380 nm) increased

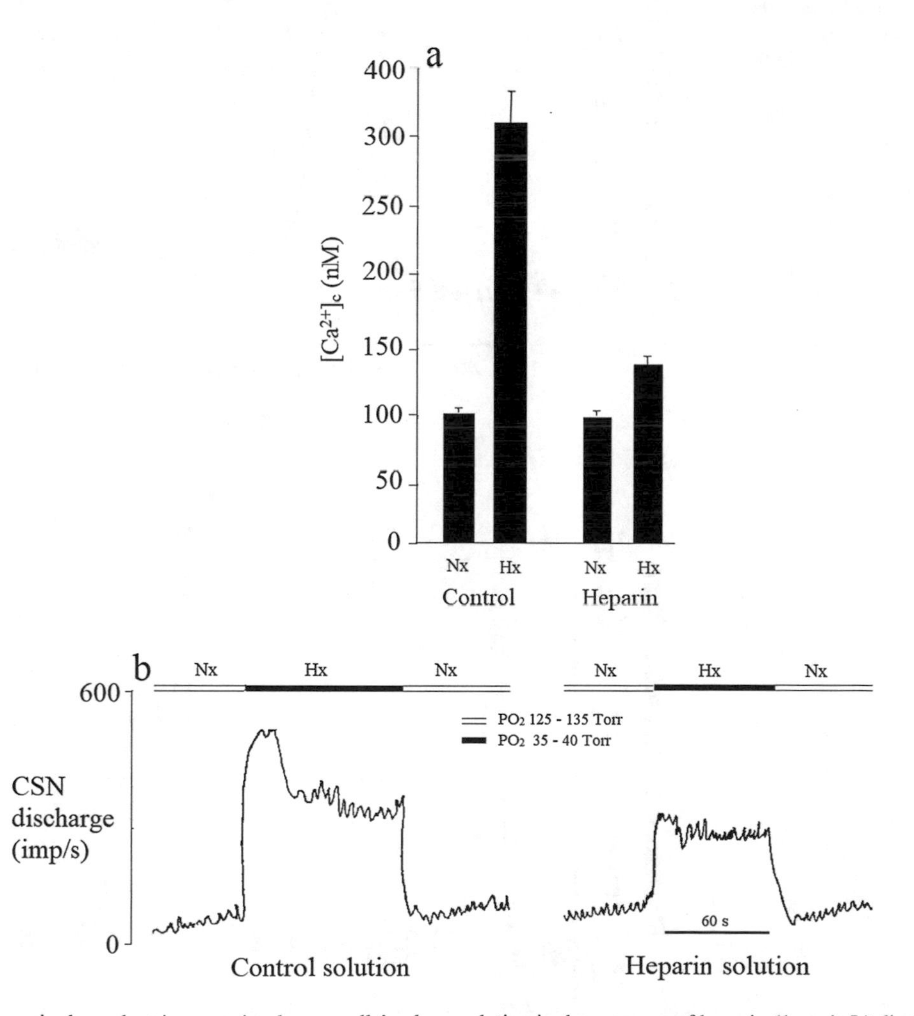

Fig. 2 Hypoxia-dependent increase in glomus cells' calcium content $[Ca^{2+}]_c$ in the control solution (n = 7) and after pretreatment of cells with heparin solution (1 mg/mL) (n = 8). Heparin diminished the hypoxia-dependent increase in $[Ca^{2+}]_c$: 347 ± 45 nM in control vs. 130 ± 6 nM in heparin solution (**a**). Likewise, perfusion with a hypoxic solution in the presence of heparin (1 mg/mL) diminished the hypoxia-dependent increase in the in vitro recorded chemosensory discharge of the carotid sinus nerve (CSN): 330 imp/s in control vs. 255 imp/s in heparin solution (**b**); *Nx* normoxia, *Hx* hypoxia. Indo-1 AM probe

from 0.71 to 1.34: a reflection of $[Ca^{2+}]_c$ increase. Replacement of hypoxic with a normoxic solution (control) promptly decreased the ratio to the initial normoxic level of 0.73. In the hypoxic test repeated in the presence of 100 μM 2-APB in the solution, a prompt rise in $[Ca^{2+}]_c$ was abrogated, which was followed by a slight increase in the fluorescence ratio (Fig. 3a). With hypercapnia (PCO_2 = 55–60 Torr, pH = 7.15), superfusion of glomus cells increased the fluorescence ratio from 0.80 present in normocapnia (PCO_2 = 30–35 Torr, pH = 7.35) to 1.65. This increase, reflecting $[Ca^{2+}]_c$ increase, was not abolished with 100 μM 2-APB (Fig. 3b). In

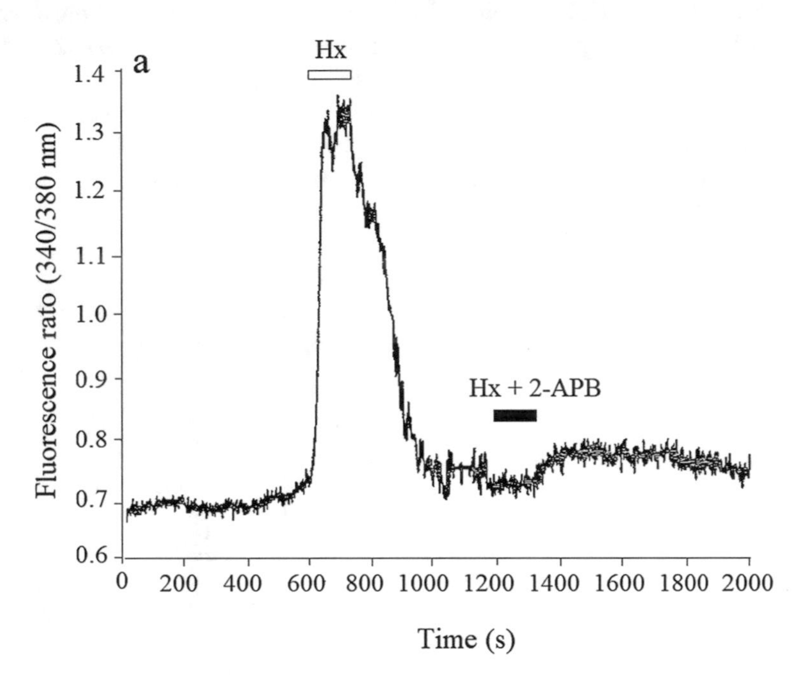

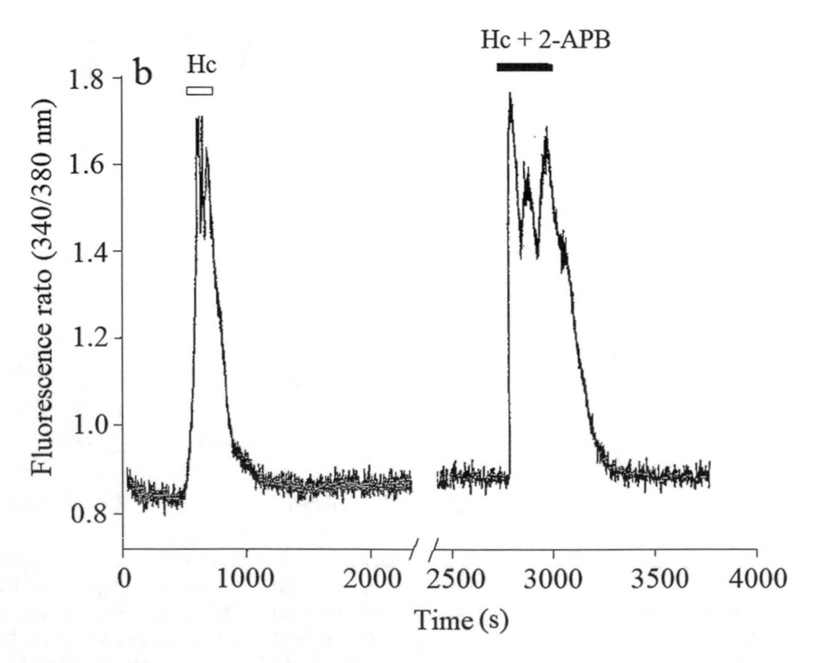

Fig. 3 Superfusion of glomus cells with a hypoxic solution (PO_2 = 12–15 Torr) increased the Fura-2 associated-fluorescence ratio from 0.72 (normoxia) to 1.35 (hypoxia); the increase was absent in the presence of 2-APB (100 μM) (**a**). A similar concentration of 2-APB failed to inhibit the hypercapnia-related increase in fluorescence ratio; the increase was from 0.80 (normocapnia) to 1.68 (hypercapnia – PCO_2 = 55 Torr, pH = 7.15) (**b**). The corollary is that $[Ca^{2+}]_c$ increase in glomus cells was not affected by hypercapnia; *Hx* hypoxia, *Hc* hypercapnia. Fura-2 AM probe

separate experiments, we used a 2.5-fold larger concentration of 2-APB in the solution, and we found inappreciable effects on $[Ca^{2+}]_c$ in glomus cells either (data not shown). These results indicate that IP_3R/s inhibition with 2-APB failed to affect the CO_2-mediated $[Ca^{2+}]_c$ augmentation in glomus cells. Thus, 2-APB produced a profound inhibitory effect in hypoxia, unlike hypercapnia.

3.4 Effect of Extracellular Ca^{2+} Chelation on Hypoxic Responses (Fura-2 AM Probe)

Change from normoxia (PO_2 = 125 Torr) to hypoxia (PO_2 = 12–15 Torr) in the superfusate containing 2.2 mM Ca^{2+} caused a repaid increase in the fluorescence ratio from 0.67 to 0.90 in glomus cells. On the other side, a change to a Ca^{2+}-free superfusate containing 1 mM EGTA abolished the rapid increase in the fluorescence ratio after the implementation of hypoxia, although a gradual exiguous increase in the ratio to 0.7 was noticeable during the following few minutes (Fig. 4). This indicates that during hypoxia, a minor part of the intracellular $[Ca^{2+}]_c$ increase was due to the store-related Ca^{2+} release. A slow and little increase in $[Ca^{2+}]_c$ in the absence of extracellular Ca^{2+} suggests that store Ca^{2+}-ATPase pump is less efficient than cell membrane operated voltage-gated receptor channels. That, however, contrasts the ATP-induced IP_3R/s activation and rapid $[Ca^{2+}]_c$ increases in the absence of extracellular Ca^{2+} (see ATP effects and explanation in Fig. 10b). A continued superfusion with Ca^{2+}-free solution

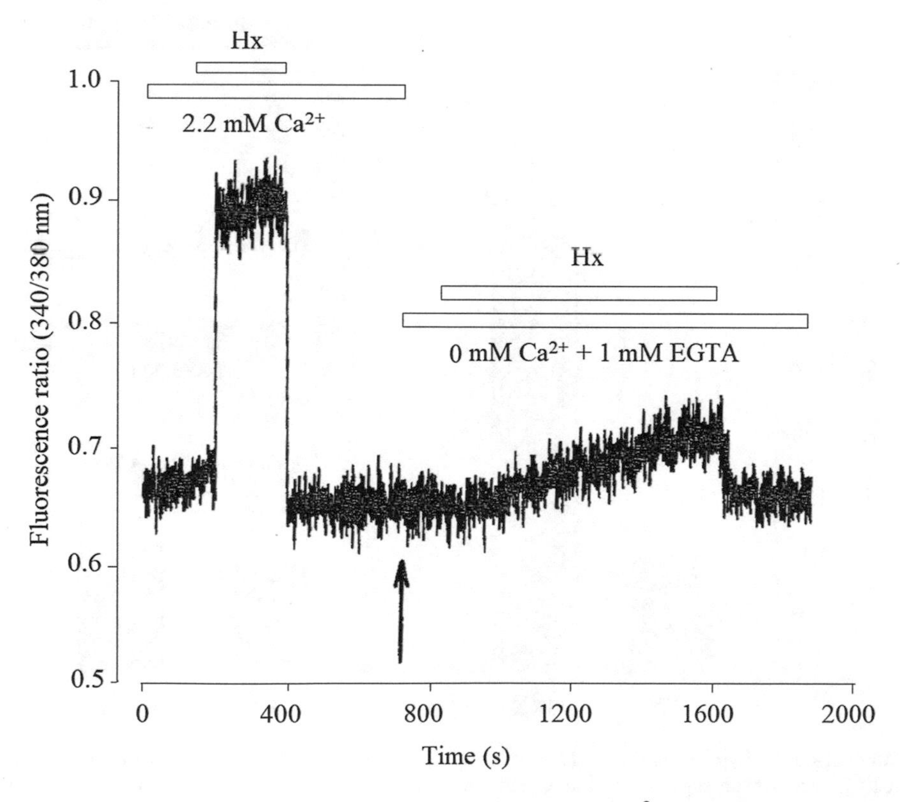

Fig. 4 Hypoxic (PO_2 = 12 Torr) superfusate increased the fluorescence ratio to 0.90 when compared to the normoxic 0.67 (PO_2 = 125 Torr) in the presence of Ca^{2+}. In the absence of Ca^{2+}, the hypoxic effect was meager and delayed as the ratio increased to 0.70. The corollary is that a minor part of $[Ca^{2+}]_c$ increase during hypoxia could be ascribed to Ca^{2+} release from intracellular stores. The arrow indicates the onset of glomus cells superfusion with EGTA, a calcium chelator; *Hx* hypoxia, *EGTA* ethylene-bis(oxyethylenenitrilo) tetraacetic acid. Fura-2 AM probe

containing 1 mM EGTA for a longer period resulted in a significant drop in the fluorescence ratio: a likely leaching effect on $[Ca^{2+}]_c$ by a chelator (Dasso et al. 1997). A resting level of Ca^{2+} declined by 50% in these cells with 1 mM EGTA in the solution for 5 min. The effect resembled the response to hypoxia in the presence of 2-APB in the solution (see Fig. 3a), which strengthens a suggestion that 2-APB blocked the Ca^{2+} influx.

3.5 Hypoxia and $[Ca^{2+}]_c$ Change in Glomus Cells: A Graded Response to 2-APB (Indo-1 AM Probe)

We tested the effects of three levels of 2-APB in hypoxic superfusate and compared the $[Ca^{2+}]_c$ increases with the control response (absence of 2-APB) in glomus cells. 2-APB, itself, did not change the $[Ca^{2+}]_c$ level during normoxic superfusion. A change from normoxic to hypoxic solution (PO_2 = 12–15 Torr) increased $[Ca^{2+}]_c$ level from 92 ± 5 nM to 345 ± 40 nM (Fig. 5 inset; n = 7). However, with a similar hypoxic solution containing 50 μM 2-APB, $[Ca^{2+}]_c$ increases were significantly less to a level of only 168 ± 19 nM. With an additional increase of 2-APB concentration in the superfusate (100 μM and 250 μM), $[Ca^{2+}]_c$ increases were nearly absent (110 ± 13 nM and 104 ± 6 nM, respectively). Since 100 μM of 2-APB nearly eliminated the hypoxia-related $[Ca^{2+}]_c$ increase, we used this concentration in the hypoxic tests on glomus cells (2–3 cells *per* view field). The mean values of $[Ca^{2+}]_c$ from 11 experiments are presented in Fig. 5. It illustrates the control

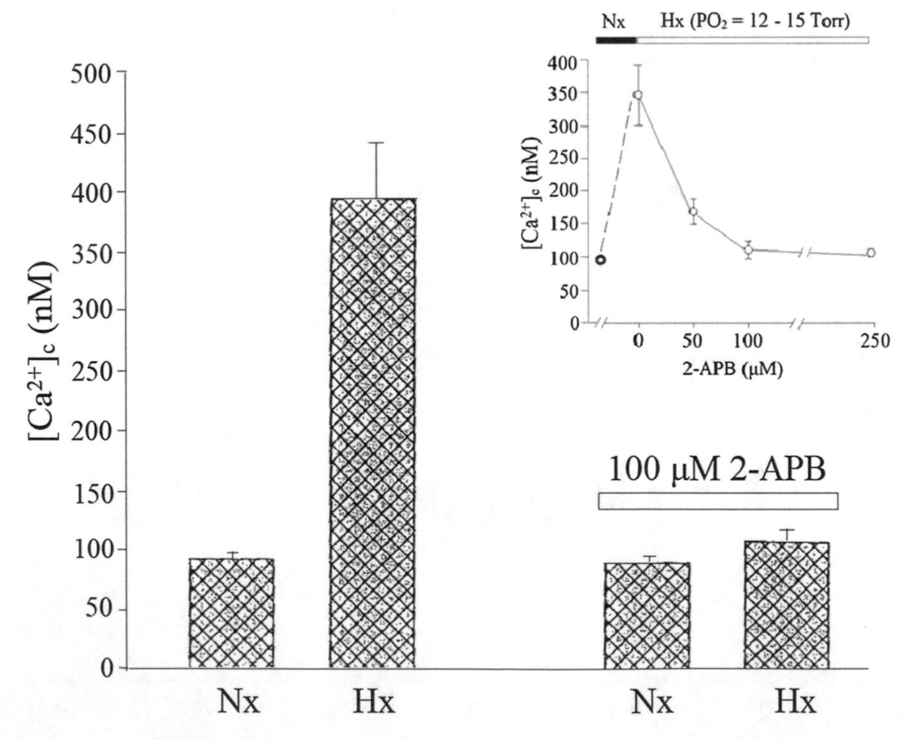

Fig. 5 Comparison of hypoxia (PO_2 = 12–15 Torr) induced $[Ca^{2+}]_c$ increases in superfused glomus cells in the absence and presence of 100 μM 2-*APB* is illustrated in a bar graph. In the control condition, hypoxia caused $[Ca^{2+}]_c$ increase from 95 ± 7 nM to 376 ± 48 nM (n = 11). Hypoxia-related $[Ca^{2+}]_c$ increases were nearly eliminated, amounting from 90 ± 6 nM to 99 ± 12 nM, in the presence of 100 μM 2-APB in the superfusate. The presence of 2-APB also failed to affect $[Ca^{2+}]_c$ level in normoxia. The inset shows the graded inhibition of $[Ca^{2+}]_c$ increases with increasing concentrations of 2-APB in the hypoxic superfusate, and the rationale for choosing the 100 μM 2-APB in further tests as this concentration abrogated $[Ca^{2+}]_c$ increases in response to hypoxia in glomus cells. Data are shown as means ±SE; *Nx* normoxia, *Hx* hypoxia. Indo-1 AM probe

hypoxic responses vs. hypoxic responses in the presence of 2-APB (100 μM). A change from normoxic to hypoxic superfusate increased $[Ca^{2+}]_c$ level from 95 ± 4 nM to 376 ± 48 nM. With 100 μM 2-APB in the superfusate, hypoxia-related $[Ca^{2+}]_c$ increases were nearly absent, suggesting a specific effect of this inhibition on hypoxia-induced $[Ca^{2+}]_c$ increase.

3.6 Hypercapnia and $[Ca^{2+}]_c$ Changes: Effect of 2-APB (Indo-1 AM Probe)

In the control condition, $[Ca^{2+}]_c$ was 100 ± 7 nM in glomus cells, and it increased to 236 ± 11 nM (n = 7) with a change in the superfusate from normocapnia (PCO_2 = 30–35 Torr, pH = 7.35–7.40) to hypercapnia (PCO_2 = 55–60 Torr, pH = 7.15–7.18). An increased concentration of 2-APB from 100 μM to 250 μM in the superfusate did not affect $[Ca^{2+}]_c$ response. This result indicates that IP_3R/s receptor inhibition with 2-APB failed to decrease CO_2-mediated $[Ca^{2+}]_c$ increases in glomus cells (Fig. 6).

3.7 Hypoxic and Hypercapnic Ca^{2+} Responses in Glomus Cells with Curcumin (Fura-2 AM Probe)

Change in the superfusing solution from normoxia and normocapnia (PO_2 = 125 Torr, PCO_2 = 30 Torr, pH = 7.32) to hypoxia (PO_2 = 13 Torr, PCO_2 = 30 Torr, pH = 7.32) increased the Fura-2-associated fluorescence ratio in glomus cells from 0.49 to 0.59, and a change to hypercapnia (PO_2 = 125 Torr, PCO_2 = 55 Torr, pH = 7.17) increased the ratio to 0.58. Repetition of similar tests in the presence of 50 μM curcumin significantly inhibited the hypoxic (0.50 to 0.54),

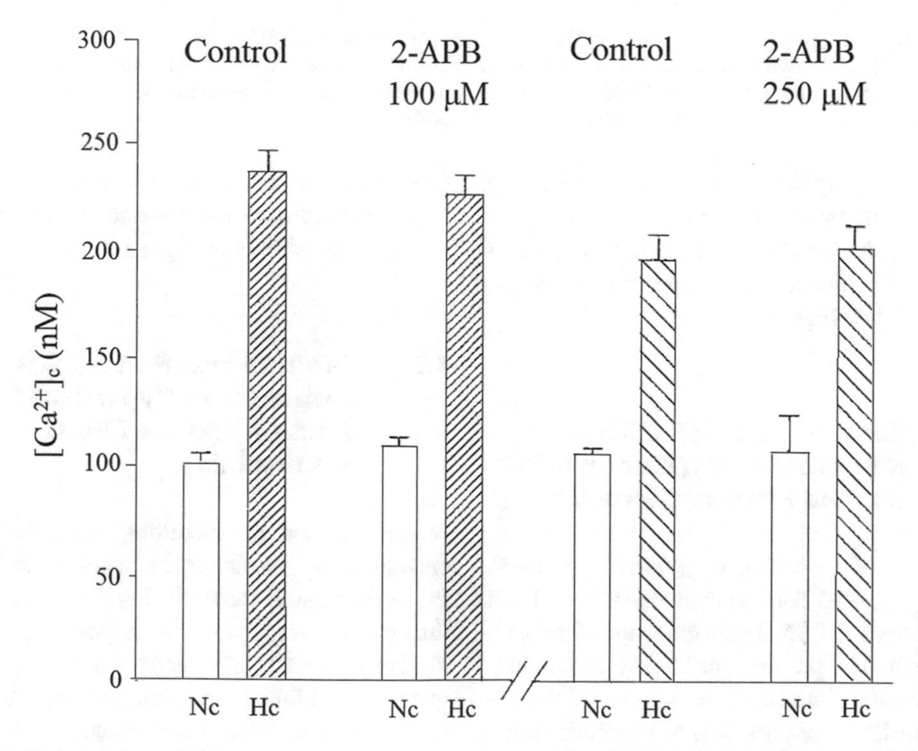

Fig. 6 Hypercapnic (PCO_2 = 55–60 Torr; pH = 7.15–7.18) solution increased $[Ca^{2+}]_c$ from 100 ± 7 nM to 236 ± 11 nM (*n* = 7) in glomus cells. Similar $[Ca^{2+}]_c$ increases were noticed in hypercapnia in the presence of 100 μm and 250 μm of 2-APB, indicating that inhibition of IP_3R/s with 2-APB did not affect $[Ca^{2+}]_c$ increase; *Nc* normocapnia, *Hc* hypercapnia. Indo-1 AM probe

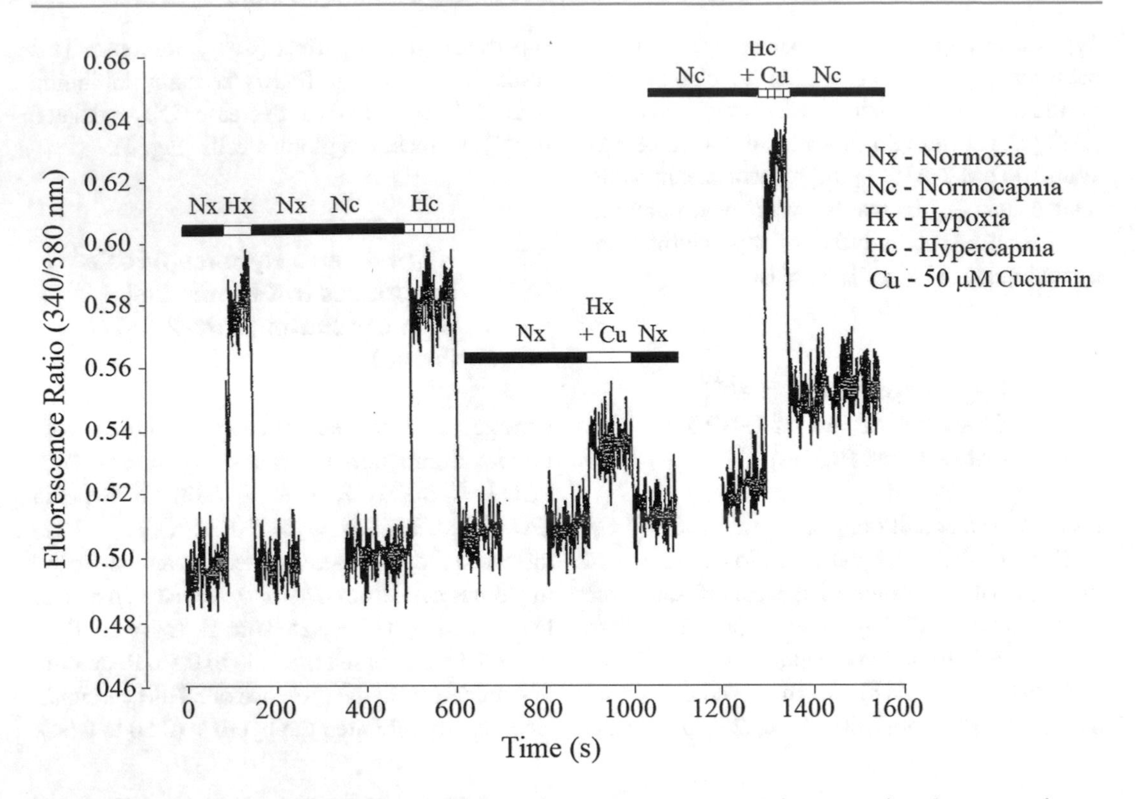

Fig. 7 Hypoxia (PO_2 = 12–15 Torr. PCO_2 = 30, pH = 7.32) and hypercapnia (PCO_2 = 55 Torr; pH = 7.15) in the superfusate increased $[Ca^{2+}]_c$ in glomus cells, which was reflected in the fluorescence ratio increases as compared to normoxic and normocapnic control. In the presence of 50 μM curcumin, the increase in hypoxia, but not in hypercapnia, was inhibited. Fura-2 AM probe

but not hypercapnic, fluorescence ratio increase (Fig. 5). This observation was akin to that with 2-APB in the solution in that it distinguished between the hypoxic and hypercapnic changes in $[Ca^{2+}]_c$ (see Figs. 3 and 7).

3.8 Carotid Sinus Nerve (CSN) Responses to Hypoxic Solution and the Effect of Curcumin

Change of perfusate from normoxia (PO_2 = 135–140 Torr) to hypoxia (PO_2 = 35–40 Torr) increased CSN discharge from 86 imp/s to 605 imp/s. The presence of 25 μM curcumin in the perfusate diminished the hypoxic CSN discharge to 490 imp/s, and 50 μM curcumin further diminished the hypoxic discharge to 175 imp/s, which was a rather drastic 71% decline of the control normoxic level. Combined results of CSN discharge in response to perfusion with a hypoxic solution in the absence and presence of curcumin are shown in Fig. 8a, b.

3.9 Carotid Sinus Nerve (CSN) Responses to Hypercapnic Solution and the Effects of Curcumin

A representative recording of the CSN chemosensory discharge in response to a high PCO_2 solution is shown in Fig. 9a. In contradistinction to responses to hypoxic perfusion, increases in CSN discharge due to high PCO_2 were not inhibited but enhanced by curcumin (Fig. 9b). In the absence of curcumin, increased PCO_2 of 50–55 Torr during normoxia stimulated CSN activity from 110 ± 8 imp/s to 248 ± 18 imp/s. In the presence of 50 μM curcumin, the

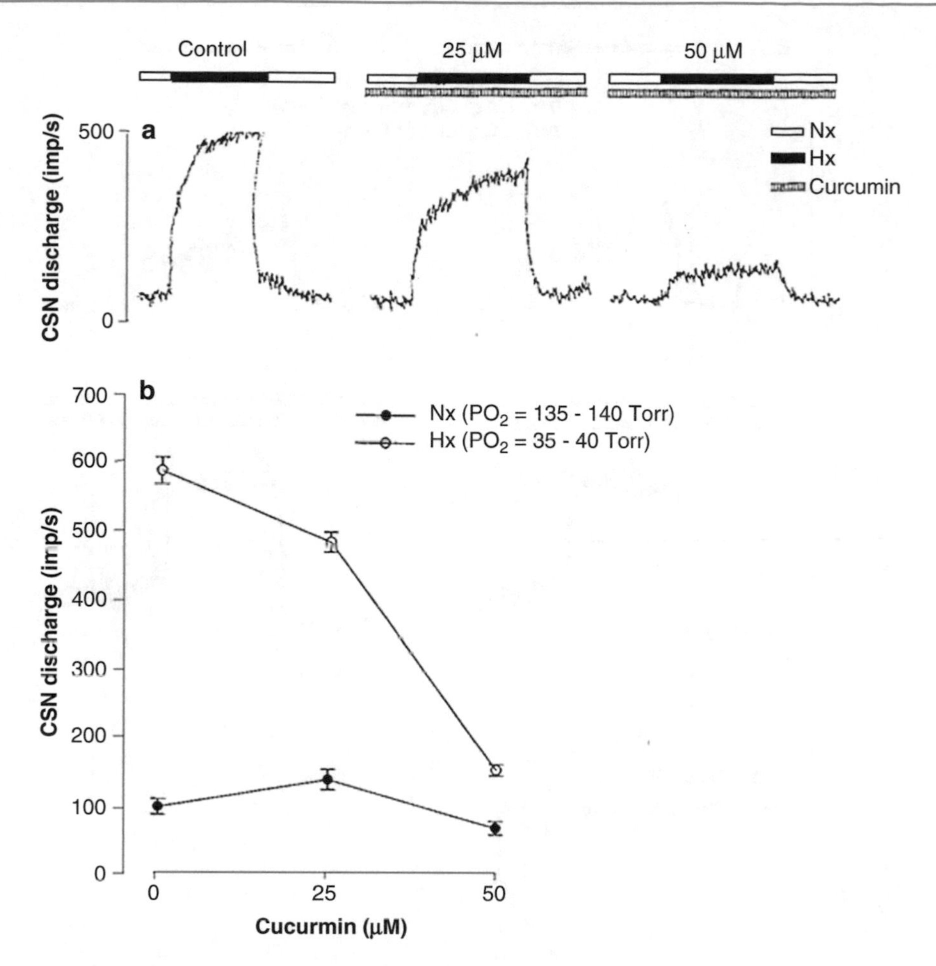

Fig. 8 Perfusion of an in vitro carotid body preparation. The hypoxic solution increased carotid sinus nerve (CSN) chemosensory discharge when compared to normoxic solution. The CSN discharge increase associated with hypoxia was partially inhibited in the presence of 25 μM curcumin in the solution. With an additional increase of curcumin to 50 μM, the increase in CSN discharge was nearly eliminated (**a**). A graph presents a summary of these results (means ±SE; n = 4) (**b**), *Nx* normoxia, *Hx* hypoxia

mean hypercapnia-related CSN discharge reached 307 ± 47 imp/s (Fig. 9c), indicating a greater sensitivity to CO_2 without a commensurate increase in $[Ca^{2+}]_c$ (see Fig. 6 for 2-APB response).

3.10 IP_3R/s Activation with ATP and Inhibition with 2-APB (Indo-1 AM and Fura-2 AM Probes)

ATP-induced $[Ca^{2+}]_c$ increases are due to Ca^{2+} influx and Ca^{2+} release from intracellular stores (Bean 1992). In APT-treated glomus cells, a two-thirds portion of $[Ca^{2+}]_c$ increase is due to Ca^{2+} influx, with the remaining part being released from intracellular stores (Mokashi et al. 2003). In the absence of extracellular Ca^{2+}, much of the $[Ca^{2+}]_c$ increase is due to the IP_3R/s activation, causing intracellular Ca^{2+} release (Levin et al. 1997). Herein we tested the effect of extracellular ATP (100 μM) in the control solution (2.2 mM Ca^{2+}) and in a Ca^{2+}-free solution that contained 1 mM EGTA. ATP increased $[Ca^{2+}]_c$ from 83 ± 12 nM to 2,100 ± 320 nM (n = 7) in the presence of Ca^{2+}. In a Ca^{2+}-free solution, $[Ca^{2+}]_c$ increases due to ATP were barely 670 ± 90 nM

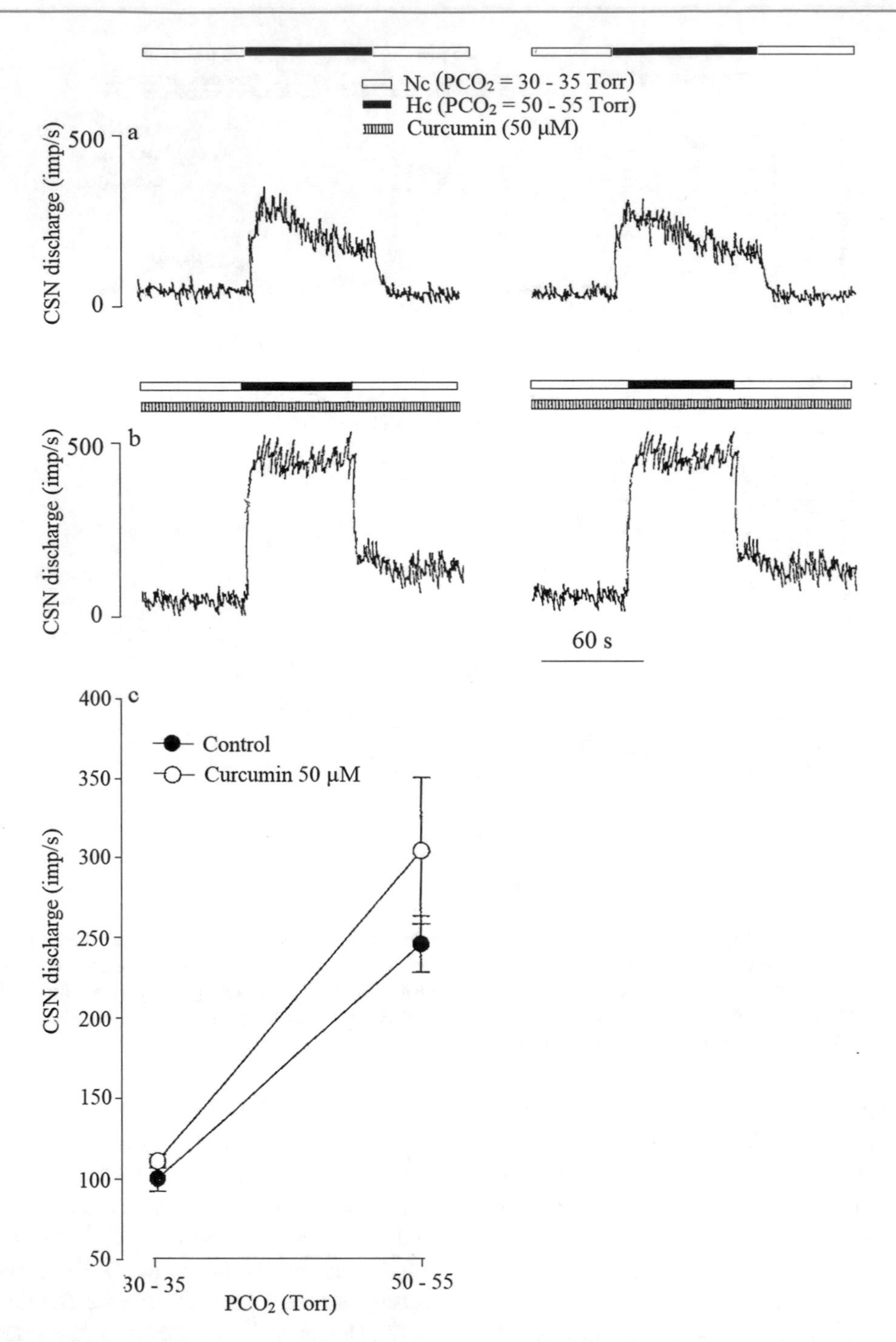

Fig. 9 Hypercapnia-associated increases in the carotid sinus nerve (CSN) discharge in the absence (**a**) and presence of 50 μM curcumin (**b**). Summary results are shown in the graph (**c**) as means ±SE (*n* = 4). Curcumin enhanced the hypercapnic CSN discharge when compared to the control response; *Nc* normocapnia, *Hc* hypercapnia

(n = 5). However, in the presence of 2-APB (100 μM), a similar concentration of ATP caused yet significantly smaller increases in $[Ca^{2+}]_c$, as it changed from 90 ± 7 nM to 142 ± 15 nM (Fig. 10a; Indo-1 AM probe). Thus, ATP-induced $[Ca^{2+}]_c$ increases via IP_3R/s activation were nearly blocked with 2-APB. Increases in $[Ca^{2+}]_c$ due to ATP and their inhibition in the

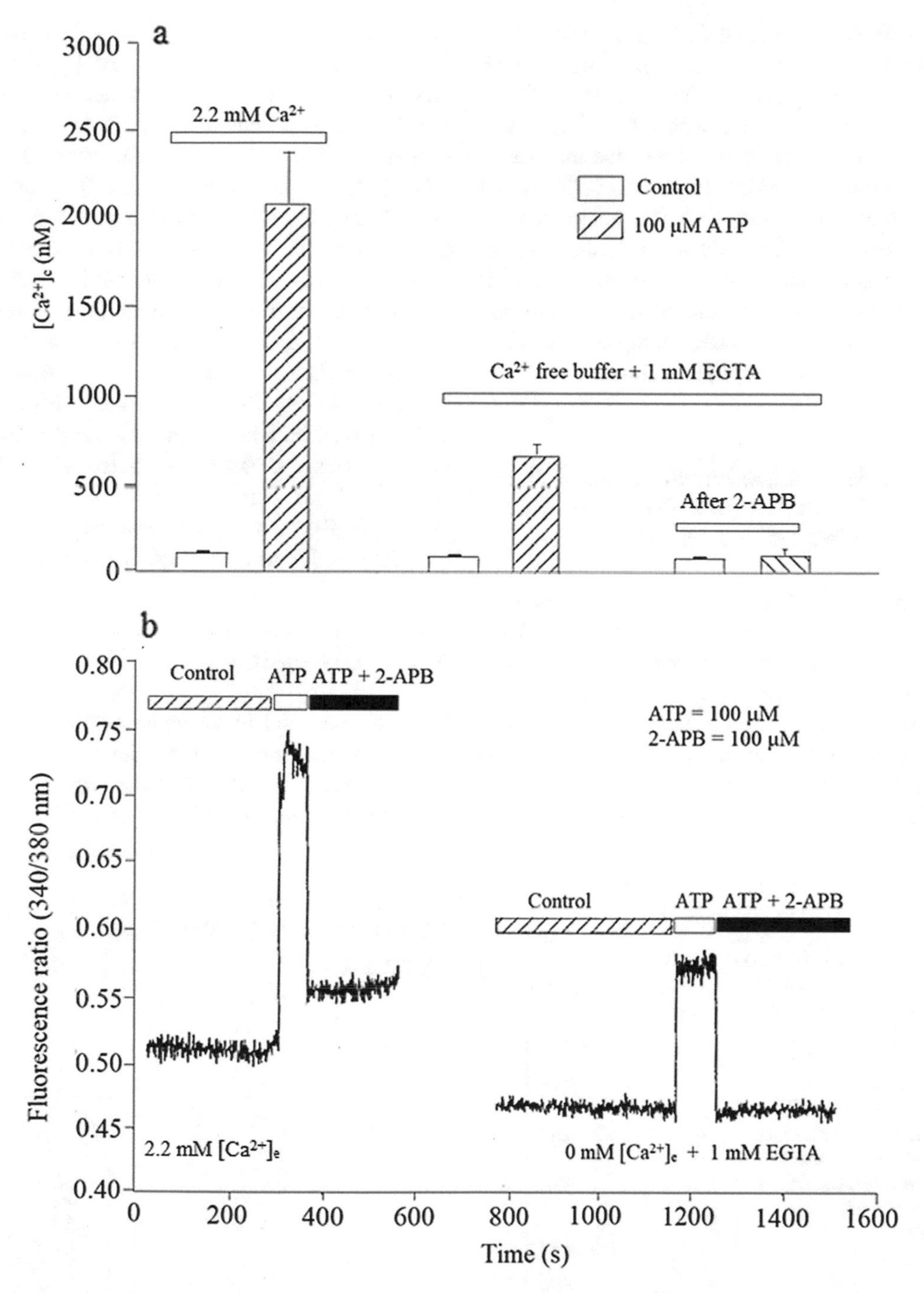

Fig. 10 ATP-induced $[Ca^{2+}]_c$ increases were tested in carotid body glomus cells during normoxic superfusion with 2.2 mM Ca^{2+} (control) and a Ca^{2+}-free solution with 1 mM EGTA. In the control, ATP (100 μM) increased $[Ca^{2+}]_c$ to 2,100 ± 320 nM ($n = 7$), whereas in the Ca^{2+}-free solution, the increase amounted to 670 ± 90 nM ($n = 5$) during normoxia. However, with a similar concentration of ATP, $[Ca^{2+}]_c$ increase was only 142 ± 12 nM in the presence of 100 μM 2-APB (Indo-1 AM probe) (**a**). A partial inhibition of ATP-associated increase in $[Ca^{2+}]_c$ with 2-APB in the presence of Ca^{2+} in solution, and total inhibition of ATP-associated increase in $[Ca^{2+}]_c$ with 2-APB in a Ca^{2+}-free solution is shown in the lower panel (Fura-2 AM probe) (**b**); *$[Ca^{2+}]_e$*, extracellular calcium content; *ATP* adenosin-5′-triphosphate, *EGTA* ethylene-bis(oxyethylenenitrilo) tetraacetic acid

presence and absence of Ca^{2+} are also illustrated in Fig. 10b (Fura-2 AM probe). It shows that with Ca^{2+} in the solution, cellular increase in $[Ca^{2+}]_c$ due to ATP was blocked with 2-APB, which was reflected by a near abrogation of the increase in the fluorescence ratio: from 0.50 to 0.73 without and to 0.55 with 2-APB (Fig. 10b, left section). In the absence of extracellular Ca^{2+}, the baseline fluorescence ratio decreased to 0.46, and it increased due to ATP-induced IP_3R/s activation to 0.57: the increase being completely blocked with 2-APB (Fig. 10b, right section).

3.11 Ryanodine Receptors (RyR/s) Activation and IP_3R/s Inhibitory Effects of 2-APB Interaction

The effects of IP_3R/s inhibition on ryanodine receptor activation were tested using caffeine (Fig. 11). This set of experiments was done to identify the possibility of 2-APB effects (IP_3R/s blocker) on the response to caffeine. Caffeine (5 mM)-induced stimulation of ryanodine receptors (RyR/s) increased $[Ca^{2+}]_c$ from 90 ± 4 nM to 370 ± 42 nM ($n = 7$) in the presence of extracellular Ca^{2+} during normoxia ($PO_2 = 125–130$ Torr). The $[Ca^{2+}]_c$ increases due to RvR/s activation were smaller when compared with ATP-induced $[Ca^{2+}]_c$ increases as shown in Fig. 10a, left part). With 2-APB (100 μM) in the solution, $[Ca^{2+}]_c$ increases (334 ± 21 nM; $n = 7$) in response to caffeine (5 mM) were fully present. The lack of changes in $[Ca^{2+}]_c$ was akin to those with 2-APB or curcumin concerning the hypercapnic response (see Figs. 3b, 6, and 7). The sustained $[Ca^{2+}]_c$ increases indicate a functional independence of the two intracellular receptor sites (IP_3R and RyR), as IP_3R inhibition with 2-APB had no effect on Ca^{2+} release channels, associated with RyR/s activation. Thus, combined effects on Ca^{2+} stores due to IP_3R/s and RyR/s were *ca* 1,000 nM (*ca* 650 +/− 370 nM; see Figs. 10a and 11).

4 Discussion

Both hypoxic and hypercapnic stimuli increase $[Ca^{2+}]_c$ in the carotid body glomus cells, and these increases stimulate CSN discharge. Calcium increases are due to the influx of Ca^{2+} through

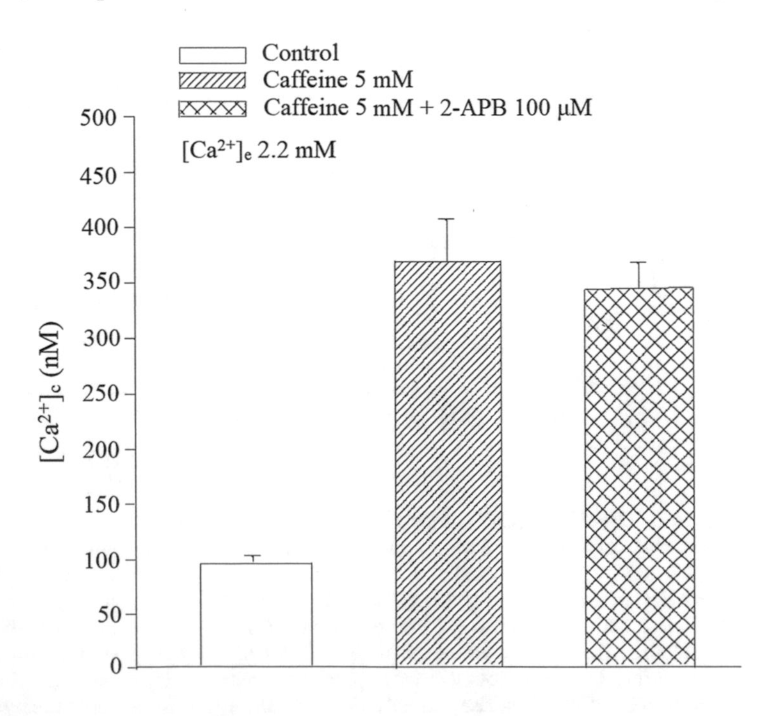

Fig. 11 Caffeine (5 mM) induced RyR/s activation and related $[Ca^{2+}]_c$ increase from 90 ± 42 nM (left-hand bar) to 370 ± 42 nM (middle bar), which was insignificantly changed to 334 ± 21 nM in the presence of 2-APB 100 μM) (right-hand bar). The result indicates that IP_3R/s inhibition with 2-APB failed to affect RyR/s activation by caffeine

membrane-associated voltage-gated and receptor-gated Ca^{2+} channels and from the intracellular store(s). We found in this study that (i) inhibition of IP_3R/s with membrane-permeant inhibitors like 2-APB and curcumin blocked hypoxia-associated increases in $[Ca^{2+}]_c$ and CSN discharge; (ii) hypercapnia-induced $[Ca^{2+}]_c$ and CSN discharge increases were not blocked with these inhibitors, which underscores functional differences in the response to hypoxic and hypercapnic stimuli; (iii) extracellular application of ATP caused $[Ca^{2+}]_c$ increase, a part of which, mainly associated with IP_3R/s, was blocked with 2-APB; and (iv) activation of RyR/s with caffeine increased $[Ca^{2+}]_c$ but 2-APB failed to inhibit that increase, which points to different Ca^{2+} signaling pathways engaged in IP_3R and RyR stimulation.

In the context of the foregoing results, two alternative hypotheses could be raised to explain the Ca^{2+} response to hypoxia and its blockade with 2-APB and curcumin: (i) influx of Ca^{2+} occurs following K^+ channel suppression and cell membrane depolarization, closely followed by Ca^{2+} depletion from the intracellular store and 2-APB blocks all the membrane depolarization, shutting off Ca^{2+} influx, and Ca^{2+} cannot be further released from the store, and (ii) hypoxia stimulates phosphatidylinositol (PI) and phospholipase C (PLC)-mediated cascade of IP_3 formation which interacts on the IP_3R/s to release Ca^{2+} from the store. There is a biological plausibility of an interaction between the two to accomplish the final result.

4.1 Role of Phosphorylation in Chemoreception

In the membrane hypothesis of CB chemosensing, hypoxia and hypercapnia result in a sequence of events which include K^+ channel inhibition to cause cell membrane depolarization, Ca^{2+} influx, and release of neuroactive substances from the glomus cells, like dopamine (DA), followed by increased CSN activity (Lahiri et al. 2001; Gonzalez et al. 1995; Peers 1990). During stimulation, transmembrane protein phosphorylation is an integral part of the intracellular signaling, associated with various kinases and phosphatases (Hunter 2000; Guillemin and Krasnow 1997). The role of PKC in the cat CB and its activation during hypoxia has been reported (Pokorski et al. 2000; Faff et al. 1999; Wang et al. 1999). Hypoxia-induced increases in lipid bilayer phosphorylation and membrane-associated phosphatidylinositol-3 kinase activation have been observed in PC12 cells (Beitner Johnson et al. 2001; Kaplin et al. 1996). Lahiri et al. (2003) and Mulligan et al. (1981) have reported that inhibition of mitochondrial phosphorylation, using oligomycin and antimycin, results in a loss of hypoxia-induced $[Ca^{2+}]_c$ increases in glomus cells and in vivo CSN discharge, but it does not affect the hypercapnic response. Thus, oxygen chemosensing and phosphorylation processes at both cell membrane and mitochondria appear to be linked phenomena in hypoxia.

In the present study, to explore the molecular mechanisms of chemosensing, we first identified the presence of IP_3R/s in glomus cells, using the IP_3R-specific antibody with IgG-conjugated FITC secondary antibody. We noticed an increase in the intensity of fluorescence in glomus cells during superfusion with a hypoxic solution when compared to the control level of fluorescence in normoxia. That finding was in line with previous studies in which enhanced fluorescence due to the IP_3R antibody has resulted from the increased formation of IP_3 and a concomitant increase in the number of IP_3R/s (Michikawa et al. 1996). We further found a significant decrease in the fluorescence caused by 2-APB binding to IP_3R/s despite the ongoing hypoxic stimulation. However, an increased level of PCO_2 made no difference to the IP_3R/s-associated fluorescence intensity in the absence or presence of 2-APB (Fig. 1a, b). These findings indicate that hypoxia is associated with IP_3R/s activation and that such a link is missing with hypercapnia as a stimulus.

4.2 IP_3R/s at the Cell Membrane, Endoplasmic Reticulum (ER), and $[Ca^{2+}]_c$ Increase

Inositol 1,4,5-triphosphate is the only known physiological activator of the cell membrane and

ER-associated IP_3R/s in smooth muscle cells and cerebral microsomes (Michikawa et al. 1996; Ehlrich et al. 1994). IP_3R/s responses leading to the activation of Ca^{2+} release channels have been reported in excitable cells (Kostyuk et al. 2000; Kaplin et al. 1996; Restrepo et al. 1992), and a topology model of IP_3R shows that it is a receptor protein that includes both voltage-gated and receptor-gated channels (Michikawa et al. 1996). Worley et al. (1987) have isolated and characterized a membrane-associated protein with high affinity binding sites for IP_3R. ER-associated localization of IP_3R in neurons has been reported by Ross et al. (1989). Thus, a close interaction of cell membrane-associated proteins, precursors of IP_3R, and ER-associated IP_3R/s during cellular stimulation is a necessary event for Ca^{2+} release. The undocking of IP_3 receptor protein from a cell membrane site and docking to cytoplasmic ER sites, is a plausible mechanism for Ca^{2+} release from the intracellular store and $[Ca^{2+}]_c$ increase on stimulation (Berridge and Irvine 1989). A similar observation of protein redocking has been made using specific antibodies in the yeast. The role of IP_3R/s-mediated intracellular Ca^{2+} release has also been identified using purified IP_3R protein in reconstituted vesicles: the release being inhibited by heparin, an IP_3R/s blocker (Ferris et al. 1990). Taken together, cellular stimulation with increased phosphorylation at the cell membrane results in the activation of IP_3R/s, their redocking to ER sites in the cytoplasm, and it further interacts with ER-MT-associated Ca^{2+} release channels (see schematic diagram in Fig. 12).

PO_2-dependence of PLC and the augmentation of ATP-mediated phosphorylation and PLC activation have been described in the hypoxic cat CB tissue (Pokorski and Strosznajder 1992, 1993, 1997). The IP_3 and diacylglycerol (DAG) molecules formed during PLC-catalyzed hydrolysis of inositol 1,4-bisphosphate (PIP_2) are conducive to $[Ca^{2+}]_c$ increases through IP_3R/s and PKC activation, respectively. However, the issue remains somewhat contentious as it has not been confirmed in the rabbit CB, possibly due to a species difference (Rigual et al. 1999). Nonetheless, the activity of classical PKC isoenzymes has been identified in the glomus cells of the cat CB, and it is augmented in hypoxia (Pokorski et al. 2000; Faff et al. 1999). PKC amplifies Ca^{2+} influx through non-voltage-operated calcium channels (Levin et al. 1997). Further, there is a crosstalk between DAG- and IP_3-mediated Ca^{2+} release (Hisatsune et al. 2005). A link between PKC

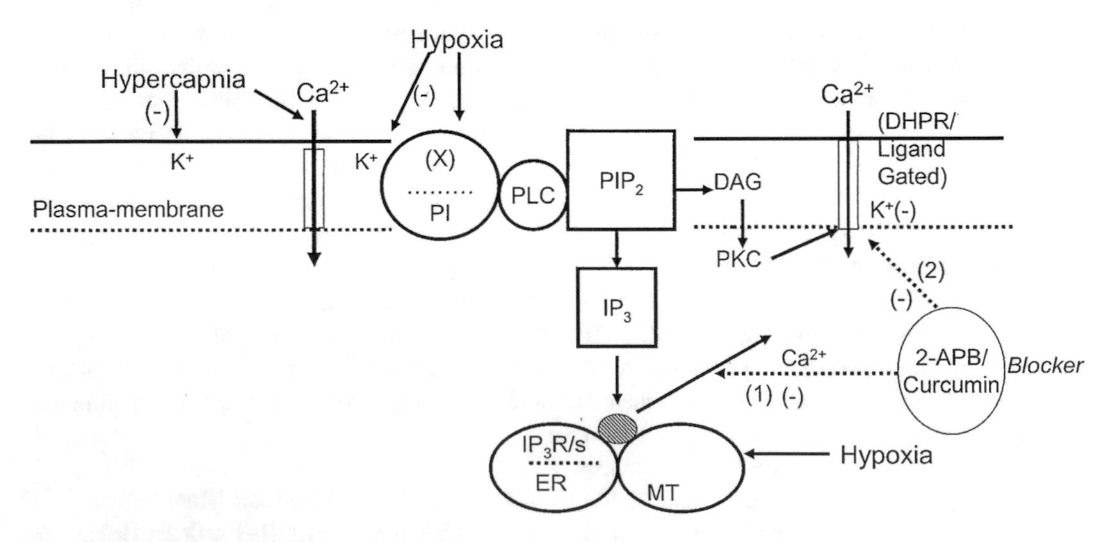

Fig. 12 Schematic diagram of chemosensing signaling pathways in carotid body glomus cells; *PLC* phospholipase C, *PIP₂* inositol 1,4-bisphosphate; *IP₃*, 1,4,5-inositol triphosphate; *DAG* diacylglycerol, *PKC* protein kinase C, *DHPR* dihydropyridine receptor, *2-APB,* 2-amino-ethoxy-diphenyl-borate; *IP_3R/s,* 1,4,5-inositol triphosphate receptors; *ER* endoplasmic reticulum, *MT* mitochondria

activity and responses of membrane K^+ channels has been reported in glomus cells in electrophysiological studies (Peers and Carpenter 1998). Also, an earlier report has indicated that PKC activation by phorbol dibutyrate ester stimulates CSN discharge in vivo during normoxia (Lahiri et al. 1990). To wrap it up, interaction of the phosphoinositide-PLC signaling pathway in shaping the molecular mechanisms of hypoxia sensing by CB seems well investigated.

4.3 IP_3R/s Activation with ATP and Inhibition with 2-APB

IP_3R/s are localized at ER but also are an integral part of cell membrane-associated proteins (Tanimura et al. 2000; Sharp et al. 1993). The presence of these two sites emphasizes the trans location of the receptor-associated proteins. Thus, using a cell membrane-permeant IP_3R/s blocker would be optimal as it can interact at the membrane and in the cytoplasm. 2-APB is one such blocker of IP_3R/s-mediated intracellular Ca^{2+} release channels (Wu et al. 2000). 2-APB acts as a membrane Ca^{2+} channel activator at low concentrations (1–5 μM), blocker at higher concentrations (>10 μM), and acts as a Ca^{2+}-ATPase pump blocker (Bilmen et al. 2002; Prakriya and Lewis 2001). Application of ATP is known to result in membrane-associated purinergic P2 receptor activation causing Ca^{2+} influx via PKC activation and increased IP_3 formation causing Ca^{2+} release from the store (Bean 1992). In the absence of extracellular Ca^{2+}, increases in $[Ca^{2+}]_c$ are possible via the IP_3R/s activation pathway alone (Levin et al. 1997). Extracellular ATP induces $[Ca^{2+}]_c$ increases in glomus cells, two-thirds of which is due to the influx of Ca^{2+} and a third due to Ca^{2+} release from the store (Mokashi et al. 2003). Herein we found that with normal Ca^{2+} containing solution (2.2 mM Ca^{2+}), ATP increased $[Ca^{2+}]_c$ and 2-APB blocked a part of that increase. However, with Ca^{2+}-free solution containing 1 mM EGTA, ATP increased $[Ca^{2+}]_c$ in the glomus cells, as expected, due to Ca^{2+} release from the ER-MT-associated stores because of only IP_3R/s activation. A repeat of similar tests with extracellular ATP in the presence of 2-APB failed to show any significant increase in $[Ca^{2+}]_c$ (Fig. 10a, b), signifying that 2-APB blocked IP_3R/s activity.

An earlier report has indicated that 2-APB interacts as a blocker of cell membrane Ca^{2+} channels (Diver et al. 2001). Hypercapnia-mediated $[Ca^{2+}]_c$ increases are primarily associated with Ca^{2+} influx (Buckler and Vaughn-Jones 1998). However, we found $[Ca^{2+}]_c$ increases due to hypercapnia both in the absence and presence of 2-APB (Figs. 3b and 6), suggesting that 2-APB may not function as membrane Ca^{2+} channel blocker. Also, the absence of any significant shift in the fluorescence ratio in the presence of 2-APB in the superfusate in normoxia indicates that IP_3R/s inhibition by 2-APB failed to affect Ca^{2+} level. That was expected since increased IP_3R/s activation is not invoked during normoxia. Bootman et al. (2002) have suggested that 2-APB acts as a blocker of store-operated Ca^{2+} entry channels, but its effects are variable concerning the inhibition of IP_3-induced Ca^{2+} release channels. These observations were based on both permeabilized and nonpermeabilized cell studies. A loss of membrane-associated component of IP_3R/s activity during membrane permeabilization may be a cause of such a mixed set of observations (Restrepo et al. 1992; Berridge and Irvine 1989). Given these conflicting observations and to confirm the role of IP_3R/s activation in hypoxia, we used curcumin as another membrane-permeant IP_3R/s blocker (Dyer et al. 2002). Curcumin is an antioxidant that readily penetrates the cytoplasm where it interacts in the organellar membrane structure (Xu et al. 1997). We noticed inhibition of hypoxia-induced $[Ca^{2+}]_c$ and CSN discharge increases and the absence of a similar type of curcumin effects in hypercapnia. Those observations were consistent with our use of 2-APB as an IP_3R/s blocker.

In the context of 2-APB and curcumin effects, it is worth mentioning the effects of heparin as a blocker of increases of IP_3R/s-related $[Ca^{2+}]_c$ in glomus cells and of CSN discharge of in vitro CB. Heparin is a charged, membrane impermeant glycosidic compound that has been used in cellular studies using a high-pressure injection technique or in membrane permeabilized procedure (Kaplin et al. 1996; Ghosh et al. 1988). We found that prolonged heparin pretreatment caused the inhibition of $[Ca^{2+}]_c$ increases in glomus cells and a partial inhibition of CSN discharge in hypoxia. That could be a result of large anionic charges on cells associated with heparin and of several other effects in CB perfusion studies. We used heparin pretreatment only, without any cell membrane permeabilization technique to avoid a possible disruption of membrane receptor sites, which would have invariably inactivated the response.

According to Ca^{2+} homeostatic mechanism, continuous stimulation of cells, particularly during hypoxia, would deplete the intracellular Ca^{2+} stores. Thus, the stores must be replenished with the influx of Ca^{2+} through the cell membrane (Kostyuk et al. 2000; Putney 1997; Berridge 1995). The notion of such a capacitative Ca^{2+} entry implies a link between the ER-MT Ca^{2+} stores and cell membrane Ca^{2+} channels. The maintenance of intracellular Ca^{2+} signaling coupled to the PLC/IP_3R/s is a complex and not well-resolved issue (Shuttleworth 1999). The function of molecular pathways engaged in the capacitative Ca^{2+} entry appears not mutually exclusive (Sweeney et al. 2002; Holda et al. 1998). Thus, inhibition of IP_3R/s with 2-APB and curcumin may deplete both the store-operated Ca^{2+} release and ligand-operated and voltage-operated membrane channels, both present in the topology model of IP3R/s, during hypoxia, rendering the store replenishment inactive. Fitzgerald (2000) has suggested that acetylcholine (ACh) is the crucial excitatory neurotransmitter in CB, based on the neurotransmitter's release during hypoxia and "synaptic-like" relationship between the glomus cells and sensory nerve endings. Moreover, ACh receptor stimulation increases $[Ca^{2+}]_c$. Muscarinic receptors appear particularly conducive to Ca^{2+} release from the store (Dasso et al. 1997), which underscores the probable role of this signaling pathway in the hypoxic response.

In this study we found that hypercapnic stimulation was unrelated to the IP_3R pathway and the inhibition of intracellular stores as a continued $[Ca^{2+}]_c$ increase was equally noticed in the absence and presence of 2-APB and curcumin. These results are in line with those obtained by Buckler and Vaughn-Jones (1994) and Roy et al. (2000) who have found that hypercapnia causes $[Ca^{2+}]_c$ increases as it depolarizes the glomus cell membrane and also increases CSN discharge. However, in contradistinction to hypoxia, hypercapnia-induced $[Ca^{2+}]_c$ increases were not inhibited by 2-APB and curcumin, the agents that interacted with IP_3R/s-mediated $[Ca^{2+}]_c$ increases at ER-MT and the receptor gated membrane. The corollary is that hypercapnic effects depend on Ca^{2+} influx through yet enigmatic mechanisms other than those known for hypoxic effects. Taken together, these results strongly support the notion that hypoxic and hypercapnic responses of carotid body glomus cells are governed by separate signaling pathways, although either having to do with $[Ca^{2+}]_c$ increases.

4.4 Ryanodine Receptor (RyR) Activation and the Role of IP_3R Inhibition

In this study, we found that caffeine-induced $[Ca^{2+}]_c$ increases were of similar magnitude in glomus cells with and without 2-APB in the solution (see Fig. 11). Thus, 2-APB failed to block RyR activation. The presence of both RyR and IP_3R receptor sites in glomus cells and their functional differences were not surprising. This finding was in line with a report by Mokashi et al. (2001) showing that RyR activation by caffeine causes $[Ca^{2+}]_c$ increase due to facilitation its release from Ca^{2+} stores, which has a rather negligible part in hypoxia-related increases in $[Ca^{2+}]_c$ and CSN discharge. The presence of separate receptor sites for IP_3 and RyR has been observed in PC12 cells and in neurons (Kaplin et al. 1996; Zaccheti et al. 1991). The functional differences of these receptors have been confirmed using the hypoxic stimulus for IP_3R activation (cyanide), IP_3R/s blocker (heparin), and RyR activation (caffeine).

4.5 Mitochondria (MT) and Membrane Link

According to the metabolic hypothesis, a significant reduction in oxidative phosphorylation reduces ATP production (Anichkov and Belenkii 1963), which leads to $[Ca^{2+}]_c$ and CSN discharge increases (Mulligan et al. 1981). Mitochondrial phosphorylation inhibitors, such as antimycin or oligomycin, block the hypoxia-induced increases without affecting the hypercapnic response (Lahiri et al. 2003). The present findings, using IP_3R/s blockers, underline the role of ER-associated Ca^{2+} release channels and receptor-gated membrane channels. Thus, oligomycin, 2-APB, and curcumin inhibit hypoxic $[Ca^{2+}]_c$ increases without affecting hypercapnic $[Ca^{2+}]_c$ increases. Taken together, these results indicate a critical role of the ER-MT complex and cell membrane, the site of IP_3R/s precursor protein, in hypoxic stimulation of CB (Lahiri et al. 2003). Functional property of mitochondria as biosensors and their role in Ca^{2+}-dependent cellular signaling has been dwelled on in previous studies (Parekh 2003; Rutter and Rizzuto 2000; Rizzuto et al. 1999). The role of mitochondria in IP_3R/s-related Ca^{2+} release channels in ER and modulation of Ca^{2+} influx also is known (Duchen 2000; Kostyuk et al. 2000). The present findings demonstrate that IP_3R/s inhibition with 2-APB and curcumin nearly blocked the hypoxic increases in glomus cell $[Ca^{2+}]_c$ and CSN discharge, without affecting the corresponding hypercapnic $[Ca^{2+}]_c$. Thus, the study confirmed the earlier hypothesis of separate CB transduction pathways of hypoxic and hypercapnic stimulation leading to the activation of CSN discharge (Mulligan et al. 1981).

Mitochondrial inhibitors decrease membrane K^+ channel currents in glomus cells (Buckler and Vaughn-Jones 1998). Inhibition of mitochondrial activity (oligomycin), IP_3R/s activity (2-APB/curcumin), and the effect on cell membrane Ca^{2+} entry channels of 2-APB (Bootman et al. 2002) suggest a close link between the ER-MT complex and membrane channels. Whether 2-APB directly interacts at the membrane site or it is an aftereffect of IP_3R/s inhibition cannot be ascertained without a systematic study on the kinetics of both receptor-gated and voltage-gated membrane ion channels in glomus cells. Receptor-gated channels may also respond rapidly. Simultaneous measurements of $[Ca^{2+}]_c$ and membrane Ca^{2+} channel activity in the absence and presence of these inhibitors are necessary to identify the sequence of events. The role of mitochondria in the control of cytosolic Ca^{2+} and its intimate connection with the ER has been extensively examined in the Hela cells (Rutter and Rizzuto 2000). Although IP_3R/s are primarily responsible for the activation of ER associated Ca^{2+} release channels, IP_3R subtype 3 has been proposed as a specific activator of cell membrane Ca^{2+} entry channels (Putney 1997; Berridge and Irvine 1989), and an IP_3R/s blocker may interact at this site as well. 2-APB/curcumin also may block voltage-gated channels that are part of IP_3R/s topology. Based on the observations herein presented, it is reasonable to sum up that (i) IP_3R/s, as well as receptor-gated and voltage-gated Ca^{2+} channels are activated at the cell membrane in hypoxia and (ii) inhibition of IP_3R/s by 2-APB/curcumin blocks Ca^{2+} release from intracellular stores and Ca^{2+} influx through receptor-gated and voltage-gated channels, despite the membrane depolarizing effects of hypoxia. It follows that the capacitative Ca^{2+} influx is blocked as the Ca^{2+} release from the store does not take place.

5 Summary and Conclusions

A summary of the effects of hypoxia and hypercapnia on the receptor sites in carotid body glomus cells is schematically illustrated in Fig. 12. The stimulatory effects of hypoxia at the cell membrane enhance receptor (X) associated phosphoinositide (PI) hydrolysis. PLC selectively catalyzes the hydrolysis of phosphatidylinositol bisphosphate (PIP_2), leading to the formation of

1,4,5-inositol triphosphate (IP_3), a water-soluble component, and diacylglycerol (DAG) that enhances protein kinase C (PKC) activity. Cell membrane stimulation inhibits membrane K^+ channels. Increased IP_3 stimulates IP_3R/s to cause Ca^{2+} release from the endoplasmic reticulum and mitochondria-associated Ca^{2+} stores. IP3R/s activation and depletion of the stores cause an additional influx of extracellular Ca^{2+} via cell membrane ligand-gated Ca^{2+} channels and dihydropyridine receptor (DHPR) – L-type voltage-sensitive channel: the influx complemented by the action of PKC. Large Ca^{2+} increases cause the neurotransmitter (s) release and related increases in CSN sensory discharge. Both 2-APB and curcumin, membrane-permeant reagents, inhibit IP_3R/s formation (1) and prevent Ca^{2+} influx through the receptor (ligand)-operated and DHPR channels (2), diminishing increases in CSN discharge in hypoxia. The inhibition of membrane channels can be a result of the direct interaction of 2-APB/curcumin on (2). During hypoxia, inhibition of receptor-associated Ca^{2+} activity also inhibits an additional Ca^{2+} influx. This chain of events signifies the importance of interaction between the cell membrane channels and the intracellular ER-MT complex in hypoxia, and it is somewhat like the capacitance hypothesis of Ca^{2+} store depletion that is followed by a related Ca^{2+} influx. Hypercapnia causes membrane depolarization to induce Ca^{2+} influx, but it is unrelated to IP_3R/s participation in the ER-MT complex since we continued to observe $[Ca^{2+}]_c$ increases in the absence and presence of 2-APB/curcumin. In effect, neurotransmitter release and CSN discharge are maintained in the presence of these blockers in hypercapnia. That also suggests that O_2-sensitive cell membrane K^+ and receptor channels respond differently when compared to CO_2-sensitive K^+ channels in the membrane.

In conclusion, hypoxia presents a global environmental issue. Herein we have tried to identify the pathways of hypoxic stimulation in the carotid body. One model describes how hypoxia caused the influx of calcium following K^+ suppression, closely followed by calcium depletion from the intracellular stores. 2-APB and curcumin block all other responses, including calcium depletion. In the absence of calcium depletion, the remaining process does not occur. The other model emphasizes the role of membrane-associated phosphorylation with increased IP_3 formation and IP_3R/s activation in hypoxia to cause Ca^{2+} release from the stores and influx, in addition to membrane depolarization. Inhibition of IP_3R/s by inhibitors blocks both release and influx events. However, hypercapnia effects on Ca^{2+} influx via membrane channels in the glomus cells are not affected by the inhibitors, which support the biological plausibility of separate pathways in O_2- and CO_2-related calcium content and carotid sinus nerve discharge increases.

Acknowledgments Supported by grants R-37-HL-43413-14, ROI-HL-50180-10, T-32-07027-29, and ONR-N-00014-01-0948. This article is a tribute to the late Professor Sukhamay Lahiri, one of the most distinguished figures of carotid body research in the second half of the twentieth century. The authors, coming from various corners of the world, had the fortune to be alumni at Prof. Lahiri's Lab at the Department of Physiology of the University of Pennsylvania in Philadelphia over the years past and to share his research passion and undaunted efforts to get to the kernels of carotid body chemosensing mechanisms.

Conflicts of Interest The authors declare no conflicts of interest in relation to this article.

Ethical Approval All applicable international, national, and/or institutional guidelines for the care and use of animals were followed. All procedures performed in studies involving animals were in accord with the ethical standards of the institutions and practice at which the studies were conducted.

References

Anichkov SV, Belenkii ML (1963) Pharmacology of the carotid body chemoreceptors. Pergamon Press, Oxford

Bean BP (1992) Pharmacology and electrophysiology of ATP activated ion channels. TIPS 13:87–90

Beitner Johnson D, Milhorn DE (1998) Hypoxia induces phosphorylation of the cyclic AMP response element binding protein by a novel signaling mechanism. J Biol Chem 273:19834–19839

Beitner Johnson D, Rust RT, Hsieh TC, Milhorn DE (2001) Hypoxia activates Akt and induces phosphorylation of GSK-3 in PC-12 cells. Cell Signal 13:23–27

Berridge MJ (1995) Capacitative calcium entry. Biochem J 32:1–11

Berridge MJ, Irvine RF (1989) Inositol phosphates and cell signaling. Nature 341:197–205

Bilmen JG, Wooton LL, Godfrey RE, Smart OS, Michelangeli F (2002) Inhibition of SERCA Ca^{2+} pumps by 2-aminoethyldiphenyl borate (2-APB) reduces both Ca^{2+} binding and phosphoryl transfer from ATP, by interfering with the pathway leading to the Ca^{2+} binding sites. Eur J Biochem 269:3678–3687

Bootman MD, Collins TJ, Mackenzie L, Roderick HL, Berridge MJ, Peppiat CM (2002) 2-aminoethoxydiphenyl borate (2-APB) is a reliable blocker of store-operated Ca^{2+} entry but an inconsistent inhibitor of $InsP_3$-induced Ca^{2+} release. FASEB J 16:1145–1150

Buckler KJ, Vaughn-Jones RD (1994) Effects of hypercapnia on membrane potential and intracellular calcium in rat carotid body type I cells. J Physiol 478 (Pt 1):157–171

Buckler KJ, Vaughn-Jones RD (1998) Effects of mitochondrial uncouplers on intracellular calcium, pH and membrane potential in rat carotid body type I cells. J Physiol 513:819–833

Conforti L, Millhorn DE (1997) Selective inhibition of a slow-inactivating voltage-dependent K^+ channel in rat PC12 cells. J Physiol 502(Pt 2):293–305

Dasso LL, Buckler KJ, Vaughan-Jones RD (1997) Muscarinic and nicotinic receptors raise intracellular Ca^{2+} levels in type I cells. J Physiol 498:327–338

De La Torre JC (1980) An improved approach to histofluorescence using SPG method for tissue monoamines. J Neurosci Methods 3:1–5

Diver JM, Sage SO, Rosado JA (2001) The inositol triphosphate receptor antagonist 2-aminoethyldiphenyl borate (2-APB) blocks Ca^{2+} entry channel in human platelets: caution for use in studying Ca^{2+} influx. Cell Calcium 30:323–329

Duchen MR (2000) Mitochondria and calcium: from cell signaling to cell death. J Physiol 529:57–68

Duchen MR, Biscoe JJ (1992) Relative mitochondrial membrane potential and $[Ca^{2+}]_i$ in type I cells isolated from rabbit carotid body. J Physiol 450:33–62

Dyer JL, Zafar Khan S, Bilmen JG, Hawtin SR, Wheatley M, Javed MH, Michelangeli F (2002) Curcumin: a new cell-permeant inhibitor of the inositol 1,4,5-triphosphate receptor. Cell Calcium 31:45–52

Ehlrich BE, Kaftan E, Bezprozvanny S, Bezprozvanny I (1994) The pharmacology of intracellular Ca^{2+} release channels. Trends Pharmacol Sci 15:145–149

Faff L, Kowaleski C, Pokorski M (1999) Protein kinase C – a potential modifier of carotid body function. Monaldi Arch Chest Dis 54:172–177

Ferris CD, Huganir RL, Snyder SH (1990) Calcium flux mediated by purified inositol 1,4,5-triphosphate receptor in reconstituted lipid vesicles is allosterically regulated by adenine nucleotide. Proc Nat Acad Sci U S A 87:2147–2151

Fitzgerald RS (2000) Oxygen and carotid body chemotransduction: the cholinergic hypothesis-a brief history and new evaluation. Respir Physiol 120:89–104

Ghosh TK, Eis PS, Millaney JM, Ebert CL, Gill DL (1988) Competitive, reversible and potent antagonism of inositol 1,4,5-triphosphate – activated calcium release by heparin. J Biol Chem 263:11075–11079

Gonzalez C, Lopez-Lopez JR, Obeso A, Perez-Garcia MT, Rocher A (1995) Cellular mechanisms in the carotid body. Respir Physiol 102:137–147

Grynkiewicz G, Poenie M, Tsien RY (1985) A new generation of Ca^{2+} indicators with greatly improved fluorescent properties. J Biol Chem 260:3440–3450

Guillemin K, Krasnow MA (1997) The hypoxic response: huffing and HIFing. Cell 89:9–12

Hain J, Onoue H, Maryleitner M, Fleischer S, Schindler H (1995) Phosphorylation modulates the function of calcium release channel of sarcoplasmic reticulum from cardiac muscle. J Biol Chem 270:2074–2081

Hisatsune C, Nakamura K, Kuroda Y, Nakamura T, Mikoshiba K (2005) Amplification of Ca^{2+} signaling by diacylglycerol-mediated inositol 1,4,5-trisphosphate production. J Biol Chem 280(12):11723–11730

Holda JR, Klishin A, Sedova M, Huser J, Blatter LA (1998) Capacitative calcium entry. NIPS 13:157–163

Hunter T (2000) Signaling-2000 and beyond. Cell 100:113–127

Kaplin AL, Snyder SH, Linden DL (1996) Reduced nicotinamide adenine dinucleotide selective stimulation of inositol 1,4,5-triphosphate receptors mediate hypoxic mobilization of calcium. J Neurosci 16:2002–2011

Kostyuk PG, Shmigol AV, Voitenko N, Svichar NV, Kostyuk EP (2000) The endoplasmic reticulum and mitochondria as elements of the mechanism of intracellular signaling in the nerve cell. Neurosci Behav Physiol 30:15–18

Kukkonen JP, Lund PE, Akerman KE (2001) 2-aminoethoxydiphenyl borate reveals heterogeneity in receptor-activated (Ca^{2+}) discharge and store-operated influx. Cell Calcium 30:117–129

Kumar GK, Overholt JL, Bright JR, Hui KY, Lu H, Prabhakar NR (1998) Release of dopamine and norepinephrine by hypoxia from PC12 cells. Am J Phys 274:1592–1600

Lahiri S, Mokashi A, Huang WX, Di Giulio C, Iturriaga R (1990) Role of protein kinase C in the carotid body signal transduction. In: Eyzaguirre C, Fidone SJ, Fitzgerald RS, Lahiri S, DM MD (eds) Arterial chemoreception. Springer-Verlag, New York

Lahiri S, Osanai S, Buerk DG, Mokashi A, Chugh DK (1996) Thapsigargin enhances carotid body chemosensory discharge in response to hypoxia in zero $[Ca^{2+}]_e$: evidence for intracellular Ca^{2+} release. Brain Res 709:141–144

Lahiri S, Rozanov C, Roy A, Storey B, Buerk DG (2001) Regulation of oxygen sensing in peripheral arterial chemoreceptors. Int J Biochem Cell Biol 33:755–774

Lahiri S, Roy A, Li J, Mokashi A, Baby SM (2003) Ca^{2+} responses to hypoxia are mediated by IP_3-R on Ca^{2+} store depletion. Adv Exp Med Biol 536:25–32

Levin R, Baiman A, Priel Z (1997) Protein kinase C induced calcium influx and sustained enhancement of ciliary beating by extracellular ATP. Cell Calcium 21:103–113

Lopez-Barneo J, Lopez-Lopez JR, Urena J, Gonzalez C (1988) Chemotransduction in the carotid body: K^+ current modulated by Po_2 in type I chemoreceptor cells. Science 241:580–582

Maruyama T, Kanaji T, Nakade S, Kanno T, Mikoshiba K (1997) 2APB, 2-aminoethoxydiphenyl borate, a membrane-penetrable modulator of Ins(l,4,5)P3-induced Ca^{2+} release. J Biochem 122:498–505

Michikawa T, Miyawaki A, Fruichi T, Mikoshiba K (1996) Inositol 1,4,5-triphosphate receptors and calcium signaling. Crit Rev Neurobiol 10:39–55

Mokashi A, Roy A, Rozanov C, Daudu P, Di Giulio C, Lahiri S (2001) Ryanodine receptor-mediated $[Ca^{2+}]_i$ release in glomus cells is independent of natural stimuli and does not participate in the chemosensory responses of the rat carotid body. Brain Res 916:32–40

Mokashi A, Li J, Roy A, Baby SM, Lahiri S (2003) ATP causes glomus cell $[Ca^{2+}]_c$ increases without corresponding increases in CSN activity. Respir Physiol Neurobiol 138:1–18

Mulligan E, Lahiri S, Storey BT (1981) Carotid body O_2 chemoreception and mitochondrial oxidative phosphorylation. J Appl Physiol 519:438–446

Parekh AB (2003) Mitochondrial regulation of intracellular Ca^{2+} signaling: more than just simple Ca^{2+} buffers. NIPS 18:252–256

Peers C (1990) Selective effect of lowered extracellular pH on Ca^{2+}-dependent K^+ currents in type I cells isolated from neonatal rat carotid body. J Physiol 422:381–395

Peers C, Carpenter F (1998) Inhibition of Ca^{2+}-dependent K^+ channels in rat carotid body type I cells by protein kinase C. J Physiol 512:743–750

Pokorski M (2000) The phosphoinositide signaling pathway in the carotid body mechanism. Bratisl Lek Listy 101(3):176

Pokorski M, Faff L (1999) Protein kinase C in the carotid body. Acta Neurobiol Exp (Wars) 59(2):159

Pokorski M, Strosznajder R (1992) Phosphoinositides and signal transduction in the cat carotid body. In: Honda Y, Miyamoto Y, Konno K, Widdicombe JG (eds) Control of breathing and its modeling perspective. Plenum, New York, pp 367–370

Pokorski M, Strosznajder R (1993) PO_2-dependence of phospholipase C in the cat carotid body. Adv Exp Med Biol 337:191–195

Pokorski M, Strosznajder R (1997) ATP activates phospholipase C in the cat carotid body in vitro. J Physiol Pharmacol 48:443–450

Pokorski M, Walski M, Matysiak Z (1996) A phospholipase C inhibitor impedes the hypoxic ventilatory response in the cat. Adv Exp Med Biol 410:397–403

Pokorski M, Sakagami H, Kondo H (2000) Classical protein kinase C and its hypoxic stimulus-induced translocation in the cat and rat carotid body. Eur Respir J 16 (3):459–463

Prakriya M, Lewis RS (2001) Potentiation and inhibition of Ca^{2+} release-activated Ca^{2+} channels by 2-aminoethyldiphenyl borate (2-APB) occurs independently of IP_3 receptors. J Physiol 536:3–19

Putney JW (1997) Type 3 inositol 1,4,5-trisphophate receptor and capacitative calcium entry. Cell Calcium 21:257–261

Restrepo D, Teeter JH, Honda E, Boyle AG, Marecek JF, Prestwich GD, Kalinoski DL (1992) Evidence for an InsP3-gated channel protein in isolated rat olfactory cilia. Am J Physiol 263:667–673

Rigual R, Cachero MT, Rocher R, Gonzalez C (1999) Hypoxia inhibits the synthesis of phosphoinositides in the rabbit carotid body. Pflugers Arch 437:839–845

Rizzuto R, Pinton P, Bin M, Chisea A, Flippin L, Pozzan T (1999) Mitochondria as biosensors of calcium microdomains. Cell Calcium 26:193–199

Ross CA, Meldolesi J, Milner TA, Satoh T, Supattapone S, Snyder SH (1989) Inositol 1,4,5-triphosphate receptor localized to endoplasmic reticulum in cerebellar Purkinje neurons. Nature 339:468–470

Roth MG (2004) Phosphoinositides in constitutive membrane traffic. Physiol Rev 84(3):699–730

Roy A, Rozanov C, Mokashi A, Lahiri S (2000) PO_2-PCO_2 stimulus interaction in $[Ca^{2+}]_i$ and CSN activity in the adult rat carotid body. Respir Physiol 122:15–26

Roy A, Li J, Al-Mehdi A, Mokashi A, Lahiri S (2002) Effect of acute hypoxia on glomus cell Em and Psi M as measured by fluorescence imaging. J Appl Physiol 93:1987–1988

Rutter GA, Rizzuto R (2000) Regulation of mitochondrial metabolism by ER Ca^{2+} release: an intimate connection. TIBS 25:215–220

Sharp AH, McPherson PS, Dawson TM, Akoi C, Campbell KP, Snyder SH (1993) Differential immunohistochemical localization of inositol 1,4,5-triphosphate and ryanodine-sensitive Ca^{2+} release channels in the rat brain. J Neurosci 13:3051–3063

Shuttleworth TJ (1999) What drives calcium entry during $[Ca^{2+}]_i$ oscillations? – challenging the capacitative model. Cell Calcium 25(3):237–246

Sweeney M, McDaniel SS, Platoshyn O, Zhang S, Yu Y, Lapp BR, Zhao Y, Thistlethwaite PA, Yuan JXJ (2002) Role of capacitative Ca^{2+} entry in bronchial contraction and remodeling. J Appl Physiol 92 (4):1594–1602

Swope SL, Moss SJ, Raymond LA, Huganir RL (1999) Regulation of ligand-gated ion channels by protein phosphorylation. Adv Second Messenger Phosphoprotein Res 33:49–78

Tanimura A, Tojyo Y, James R (2000) Evidence that type I, II, and III inositol 1,4,5 triphosphate receptors

can occur as integral cell membrane proteins. J Biol Chem 35:27488–27493
Thompson CH, Kemp CJ, Radda GK (1992) Changes in high energy phosphates in rat skeletal muscle during acute respiratory acidosis. Acta Physiol Scand 146:15–19
Vicario I, Obeso A, Rocher A, Lopez-Barneo J, Gonzalez C (2000) Intracellular Ca^{2+} stores in chemoreceptor cells of the rabbit carotid body: significance for chemoreception. Am J Phys 279:51–61
Wang ZZ, He L, Dinger CJ, Stensas L, Fidone S (1999) Protein phosphorylation signaling mechanism in carotid body chemoreception. Biol Signals Recept 8:366–374
Worley PF, Baraban JM, Supattapone S, Wilson VS, Snyder SH (1987) Characterization of inositol triphosphate receptor binding in brain. J Biol Chem 262:12132–12140
Wu J, Kamimura N, Takeo T, Suga S, Wakui M, Maruyama T, Mikoshiba K (2000) 2-aminoethoxydiphenyl borate modulates kinetics of intracellular Ca^{2+} signals mediated by inositol 1,4,5-triphosphate-sensitive Ca^{2+} stores in single pancreatic acinar cells of mouse. Mol Pharmacol 58:1368–1374
Wyatt CN, Buckler KJ (2004) The effect of mitochondrial inhibitors on membrane currents in isolated neonatal rat carotid body type I cells. J Physiol 556:175–191
Xu YX, Pindolia KR, Janakiraman N, Noth CJ, Chapman RA, Gautam SC (1997) Curcumin, a compound with anti-inflammatory and antioxidant properties, downregulates chemokine expression in bone marrow stromal cells. Exp Hematol 25:413–422
Yoshioko H, Miyake H, Smith DS, Chance B, Sawada T, Nioka S (1995) Effect of hypercapnia on ECOG and oxidative metabolism in neonatal dog brain. J Appl Physiol 78:272–2278
Zaccheti D, Cleminti E, Fasoloto C, Lorenzon P, Zottin M, Govaz F, Fumagalli G, Pozzan T, Meldolesi J (1991) Intracellular Ca^{2+} pools in PC-12 cells. J Biol Chem 266:20152–20158

Adv Exp Med Biol - Clinical and Experimental Biomedicine (2021) 11: 27–35
https://doi.org/10.1007/5584_2020_568

Published online: 22 July 2020

The Role of Hyperbaric Oxygen Treatment for COVID-19: A Review

Matteo Paganini, Gerardo Bosco, Filippo A. G. Perozzo, Eva Kohlscheen, Regina Sonda, Franco Bassetto, Giacomo Garetto, Enrico M. Camporesi, and Stephen R. Thom

Abstract

The recent coronavirus disease 2019 (COVID-19) pandemic produced high and excessive demands for hospitalizations and equipment with depletion of critical care resources. The results of these extreme therapeutic efforts have been sobering. Further, we are months away from a robust vaccination effort, and current therapies provide limited clinical relief. Therefore, several empirical oxygenation support initiatives have been initiated with intermittent hyperbaric oxygen (HBO) therapy to overcome the unrelenting and progressive hypoxemia during maximum ventilator support in intubated patients, despite high FiO2. Overall, few patients have been successfully treated in different locations across the globe. More recently, less severe patients at the edge of impending hypoxemia were exposed to HBO preventing intubation and obtaining the rapid resolution of symptoms. The few case descriptions indicate large variability in protocols and exposure frequency. This summary illustrates the biological mechanisms of action of increased O_2 pressure, hoping to clarify more appropriate protocols and more useful application of HBO in COVID-19 treatment.

Keywords

COVID-19 · Hyperbaric oxygen therapy · Hypoxemia · Inflammation

M. Paganini and G. Bosco (✉)
Department of Biomedical Sciences, University of Padova, Padova, Italy
e-mail: gerardo.bosco@unipd.it

F. A. G. Perozzo, E. Kohlscheen, R. Sonda, and F. Bassetto
Plastic and Reconstructive Surgery Unit, Padova University Hospital, Padova, Italy

G. Garetto
ATIP Hyperbaric Medical Center, Padova, Italy

E. M. Camporesi
Teamhealth Anesthesia Attending, Emeritus Professor of Surgery, USA, Tampa, FL, USA

S. R. Thom
Emergency Medicine, University of Maryland, Baltimore, MD, USA

1 Introduction

Coronavirus disease 2019 (COVID-19) is a health emergency that is saturating the care and receptive capacities of many national health systems. While most of the patients (up to 81% of the total) do not show any symptom or present with flu-like illness, others can develop severe respiratory compromise and must be hospitalize due to interstitial pneumonia with consequent hypoxia (Wu and McGoogan 2020). Besides, recent works suggest that most severe cases are characterized by a complex pattern of systemic activation consequent to cytokine storm resulting in immune system impairment and pro-inflammatory imbalance (Mehta et al. 2020).

Given the significant mortality and morbidity associated with COVID-19, the beneficial potential of adjunctive therapies cannot be dismissed. Several treatments, such as antiviral, antimalarial, or immunosuppressant drugs, are tested, as well as hyperbaric oxygen (HBO). To date, evidence of the clinical utility of HBO in COVID-19 is still limited (De Maio and Hightower 2020; Moon and Weaver 2020), but the interest is growing, and at least three trials have been registered online (ChiCTR 2020; US National Library of Medicine 2020). The objective of this article is to review and discuss HBO mechanisms of action and data addressing possible benefits, adverse effects, and potential applications in treating COVID-19 patients.

2 Hyperbaric Oxygen (HBO): Mechanisms of Action

HBO therapy is based on the laws of gas physics related to pressure and involves the intermittent inhalation of 100% oxygen in pressurized chambers. Most studies have involved oxygen administration between 1.5 and 3.0 atmosphere absolute (ATA), a range in which risks of adverse effects are minimized while obtaining therapeutic effects.

HBO increases the partial pressure of oxygen in plasma and tissues (Camporesi and Bosco 2014) and is commonly used in the treatment of decompression sickness, carbon monoxide intoxication, arterial gas embolism, necrotizing soft tissue infections, chronic skin ulcers, severe multiple trauma with ischemia, and ischemic diabetic foot ulcers (Moon 2019; Thom et al. 2011b). The differences and advantages of HBO therapy from atmospheric oxygen absorption are the following: (a) the improvement in diffusion efficiency of oxygen through the alveolar barrier; (b) the higher physically dissolved oxygen content in the blood, more than the combined hemoglobin transport capacity; and (c) the increased diffusion distance of oxygen. Altogether, these properties meet the demand of aerobic metabolism in hypoperfused regions of the body.

3 Hyperbaric Oxygen (HBO) and Inflammation

HBO has beneficial effects in reducing the inflammatory state by modulating oxidative stress, including lipid peroxidation, and increasing antioxidant enzymes (Thom 2011; Bosco et al. 2007). Accordingly, in animal models, HBO can modulate the inflammatory response and cytokine level (Halbach et al. 2019; Pedoto et al. 2003) or reduce TNF-α production and lung neutrophil sequestration (Yang et al. 2001). Studies in humans have confirmed this experimental evidence concerning the benefit emanating from HBO during different inflammatory states (Bosco et al. 2018; Marmo et al. 2017; Li et al. 2011). In the diabetic patient, whose peripheral arterial vascularization is compromised, HBO is indicated for its ability to increase tissue oxygenation and limit ischemic damage through several mechanisms (Thom et al. 2011a). In the same vein, HBO can improve the perfusion of peripheral systems by reducing the risk of multiple organ failure (MOF) (Bosco et al. 2014; Rinaldi et al. 2011; Yang et al. 2006).

Oxidative stress and reactive species of oxygen (ROS) and nitrogen have complex effects on cell signaling mediators such as HIF-1α and NK-kB. There are competing pathways that influence levels of these agents in cells, and some evidence exists for a beneficial influence of HBO in certain situations (Bosco et al. 2018). The HIF-1α and NF-kB cross talk regulates essential inflammatory functions in myeloid cells. HIF-1α increases macrophage aggregation, invasion, and motility. Whereas HIF-1α drives the expression of pro-inflammatory cytokines, enhanced microbial clearance can limit the overall production of inflammatory mediators. HIF-1α enhances intracellular bacterial killing by macrophages and promotes granule protease production and release of nitric oxide (NO) and TNF-α, which in turn further contribute to antimicrobial control. HIF-1α in myeloid cells increases the transcription of key glycolytic enzymes, resulting in increased glucose uptake and glycolytic rate. HBO can increase HIF-1α via an oxidative stress response mediated in part by thioredoxin to increase the recruitment of stem cells. HBO can also increase the activity of iNOS in leukocytes and eNOS in platelets (Thom et al. 2006, 2008, 2011a, 2012).

4 Coronavirus Disease 2019 (COVID-19) Pathogenesis: Between Inflammation and Cytokine Storm

To date, there is no specific antiviral medicine shown to be effective in preventing or treating COVID-19. Those suffering severe illness appear to have higher initial viral load and prolonged viral shedding suggestive of a failure to clear the infection due to an inadequate immune response (Liu et al. 2020b). Based on murine studies and viral shedding and IgG production patterns present in the past severe acute respiratory syndrome (SARS) outbreak due to SARS-CoV-1, lung injuries may arise due to an excessive or aberrant host inflammatory response.

Severe COVID-19 manifests clinically as acute lung injury associated with high initial virus titers, macrophage/neutrophil accumulation in the lungs, and elevated pro-inflammatory serum cytokines (Conti et al. 2020; Kowalewski et al. 2020). During infections, pathogens first encounter the innate immune system that directs anti-pathogen effects and induces adaptive immune responses. One innate inflammatory pathway involves the inflammasome, a multimeric protein complex that is responsible for the activation of caspase-1. Caspase-1, in turn, processes members of the IL-1 family of cytokines into their active forms leading to their secretion. These cytokines, including IL-1α, IL-1β, and IL-18, are pro-inflammatory and may induce either protective or damaging host response. IL-1 and IL-18 participate in the control of viral replication. IL-18, through its ability to increase interferon-gamma, seems to have a strong host pro-survival effect in coronavirus infections (Liu et al. 2020a).

IL-1α and IL-1β induce recruitment of neutrophils, polarize T-cells, and promote dendritic cell activation for priming. However, IL-1β overproduction is linked to a wide range of inflammatory pathologies, including those caused by respiratory viruses (Kim et al. 2015). IL-1β plays a central role in many inflammatory responses because of its auto-catalytic production and because it can trigger the synthesis of alternative cytokines and other inflammatory agents. Several coronavirus accessory proteins and the envelope (E) protein trigger robust activation of the inflammasome NOD-, LRR-, and pyrin domain-containing protein 3 (NLRP3). The E protein is involved in virulence and specifically correlates with enhanced pulmonary damage, edema accumulation, and death. This protein establishes an ion channel in host cells to induce NLRP3 inflammasome activation resulting in overproduction of IL-1β. The central issue concerning possible effects of HBO pertains to the pathological role of inflammasome activation and, specifically, the role of IL-1β (Debuc and Smadja 2020).

5 Hyperbaric Oxygen Therapy (HBO) and Coronavirus Disease 2019 (COVID-19)

From the clinical standpoint, prediction of arterial oxygenation at increased atmospheric pressure in patients with pulmonary gas exchange impairment can be extrapolated from published clinical data and the application of gas laws (Moon et al. 1987). A few published case series of HBO-treated COVID-19 patients appear to follow similar ratios, reporting improved survival (Guo et al. 2020; UHMS 2020) and success in preventing mechanical ventilation (Thibodeaux et al. 2020). Yet it is still unclear whether the clinical course of those patients, when improved, was due to HBO itself or only time. In the following sections, we will detail possible positive and negative interactions between HBO and COVID-19, as depicted in Fig. 1. According to the

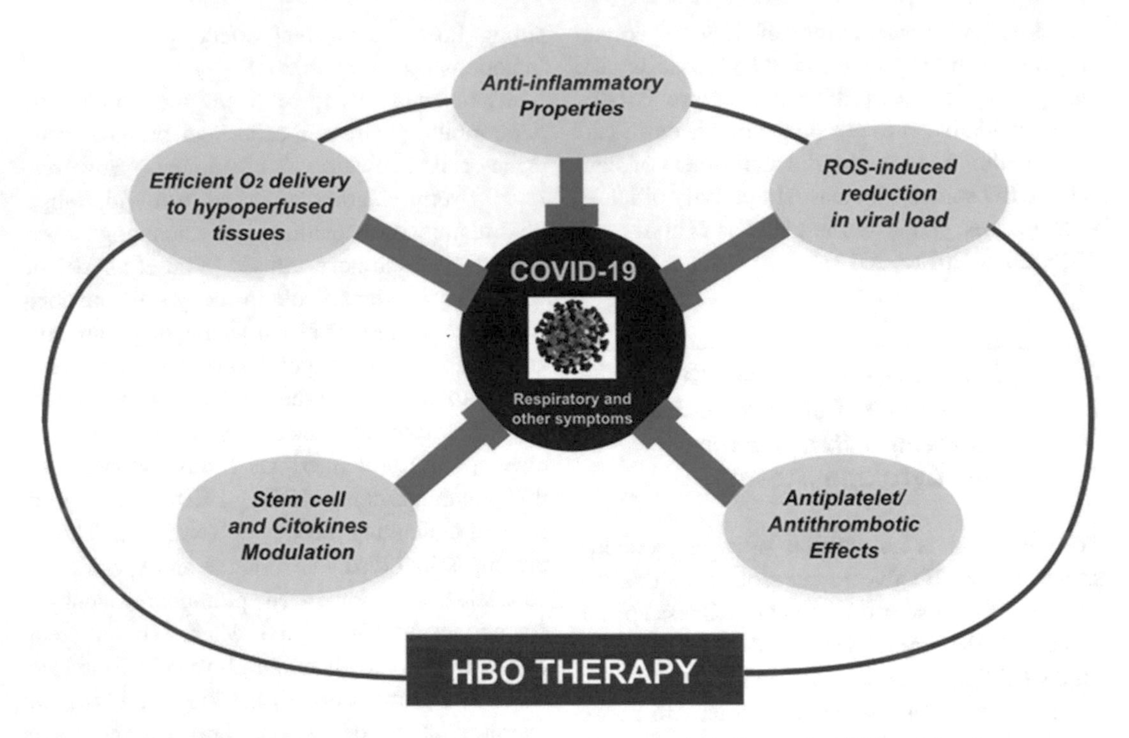

Fig. 1 Possible effects of hyperbaric oxygen (HBO) therapy in coronavirus disease 19 (COVID-19) patients

available literature, the increased amount of oxygen in the plasma could mobilize stem cells, block the inflammatory cascade, interfere with interstitial fibrosis development in the lungs, delay the onset of severe interstitial pneumonia, and reduce the risk of multiple organ failure (MOF) due to an overall abated COVID-19 viral load. However, all these possible effects have yet to be demonstrated and should be weighed on possible harms deriving from the administration of HBO.

6 Hyperbaric Oxygen Therapy (HBO) and Viral Diseases

With data showing the anti-inflammatory potential of HBO, questions have been raised of whether HBO may serve as an adjunct antiviral treatment in patients with viral pneumonia. However, interactions between HBO and viral infections are still poorly understood. Several studies have shown that oxidative stress can play a role in the progression of the HIV disease (Baugh 2000). It has also been suggested that oxidative stress can contribute to increased viral replication, transcription, or reactivation of latent infection (Peterhans 1997; Pace and Leaf 1995). Nevertheless, previous studies regarding HBO and viral replication reported uncertain effects (Hosokawa et al. 2014; Peng et al. 2012; Savva-Bordalo et al. 2012; Wong et al. 2008; Gabrilovich et al. 1990).

There is consensus in the literature that a viral infection per se does not trigger oxidative stress, while it is the host defense armamentarium that induces ROS to counter viral effects. Viruses use phospholipids and proteins taken from the host membranes to make their capsid (envelop), and ROS avidly react with phospholipids, modifying their structure, thus function. The rationale in using HBO in viral infections could thus be linked to the increased production of ROS. Since a higher viral load is associated with a more severe manifestation of COVID-19, it could be reasonable to test HBO to hamper viral replication and possibly to reduce the viral load, firstly on cultured cells and then on patients presenting with a moderate but potentially evolving disease.

7 Hyperbaric Oxygen Therapy (HBO), Cytokine Storm, and Stem Cells

Recently, intermittent hyperoxia or different oxygen partial pressures demonstrated to have an impact on stem cell proliferation, cytokine expression, and neuroprotection (MacLaughlin et al. 2019; Schulze et al. 2017; Milovanova et al. 2009). Also, Gardin et al. (2020) have suggested that the exposure of mesenchymal stem cells to HBO in in vitro simulated inflammatory conditions with pro-osteogenic factors increases the differentiation toward the osteogenic phenotype. In severe COVID-19, release of IL-1 beta and IL-6 seems to sustain a pro-inflammatory state, predisposing to pulmonary fibrosis (Conti et al. 2020; Kowalewski et al. 2020). Specifically, IL-6 and IL-8 have been associated with faster progression of pulmonary fibrosis, while IL-10, TGF-beta, IL-4, and IL-13 have not shown statistically significant associations (Papiris et al. 2018). In such an environment, HBO could attenuate the production of pro-inflammatory cytokines, as already described in response to post-surgery inflammation (Bosco et al. 2007) and in modulation of immune responses (Thom 2011). More specifically, a single preoperative HBO session the day before pancreatic surgery demonstrated to alter the inflammatory response, decreasing the pro-inflammatory IL-6 and increasing the anti-inflammatory IL-10 (Bosco et al. 2014). Also, a course of HBO treatments determined a significant reduction in TNF-α and IL-6 plasma levels in patients with avascular femoral necrosis (Bosco et al. 2018). Therefore, HBO could be used in COVID-19 to reduce cytokine levels and to enhance mobilization of stem cells from the bone marrow, especially mesenchymal stem cells, to the most damaged sites (Debuc and Smadja 2020).

8 Hyperbaric Oxygen Therapy (HBO) and Other Possible Effects in Coronavirus Disease 19 (COVID-19)

Nitrogen oxide has multiple and fundamental capacities, including vasodilation, reduced platelet activation, and decreased leukocyte adhesion to the endothelium and consequent diapedesis (Bosco et al. 2010; Thom 2009). Breathing oxygen underwater (12-m depths) – a condition like HBO therapy – showed to improve the antioxidant activity of lymphocytes and to preserve calcium homeostasis, suggesting a protective role in the physiological functions of lymphocytic cells (Morabito et al. 2011). Overall, these effects of HBO could counteract several modifications recently demonstrated in COVID-19 patients, such as disturbances of aggregation (Ciceri et al. 2020; Xu et al. 2020a) and coagulation (Lodigiani et al. 2020), and immune system impairment (Giamarellos-Bourboulis et al. 2020). For example, daily treatments with HBO could reduce platelet activation and aggregation in the lungs, thus hampering the development of pulmonary microcirculation dysfunction and catastrophic inflammation already reported in autoptic and pathologic studies (Luo et al. 2020; Xu et al. 2020b).

9 Possible Adverse Effects of Hyperbaric Oxygen Therapy (HBO)

Conventional HBO therapy has several drawbacks. From the logistical standpoint, dedicated large-scale equipment and complex structure are needed to administer HBO. Given the high infectivity of SARS-CoV-2, disinfection measures should be further strengthened, and the attending personnel should follow medical protection guidelines of the European Committee for Hyperbaric Medicine (ECHM) and the European Underwater and Baromedical Society (EUBS) (ECHM- EUBS 2020; UHMS 2020).

The main side effects of HBO are limited to the pulmonary and neurological areas. Pulmonary toxicity usually manifests with tracheobronchial irritation. Oxygen toxicity has been previously reported (Heyboer et al. 2017; Clark et al. 1999; Thorsen et al. 1998; Clark and Lambertsen 1971), but current protocols below 3 ATA minimize this risk. Randomized studies in animal models report damages to the complex phospholipidic system of the lung surfactant, with a consequent increase in alveolar surface tension causing atelectasis and hyperoxic toxicity (Webb et al. 1966). Changes in protein and phospholipid complexes in the surfactant after prolonged periods of oxygen exposure have also been reported (Prokof'ev et al. 1995; Bergren and Beckman 1975). Treatment protocols aimed at limiting O_2 exposure at high ATA have shown that the effect on the surfactant of such exposures is of no clinical relevance (Fife and Piantadosi 1991). Further, it is now consolidated that HBO does not compromise lung function in patients without chronic lung disease (Hadanny et al. 2019), but specific effects on COVID-19 patients are unknown.

Clinical signs of neurological toxicity include visual impairment, tinnitus, nausea, facial spasms, dizziness, and disorientation. In the worst-case scenario, seizures and loss of consciousness may develop, which can be rapidly treated with removal of the oxygen mask and restoration of atmospheric pressure. More recent hyperbaric therapy protocols have allowed overcoming or minimizing these issues, mainly through air respiration pauses, reduction of each session duration (< 2 h), and the use of pressures below the threshold of neural toxicity (US Navy 2008). Other effects, such as those reported on the crystalline lens, are usually reversible over time (Anderson and Farmer 1978). Additionally, adverse effects of HBO seem of smaller concern in case of COVID-19 in the face of rather limited number of patients in whom such therapy could be considered and applied.

10 Conclusions

HBO therapy demonstrates multiple beneficial effects and rare, but preventable, adverse consequences. Since the pathophysiology of COVID-19 has not yet been clarified, several questions about the potential clinical utility of HBO in treating this infection remain unanswered. We believe that the literature findings summarized in this article provide the practitioners with the necessary evidence to consider HBO as adjunctive treatment for COVID-19. Further, patients should be treated carefully, in hospital-based facilities that can manage the patient transport issues and can provide adequate infection control strategies to ensure safety of healthcare personnel. Through careful monitoring of the patient during the HBO treatment, major side effects could be early detected and avoided. Nevertheless, HBO therapy should be considered carefully, after weighing harms and potential benefits, and it should be tailored to the patient and beforehand verified through a rigorous scientific process.

Acknowledgments We would like to thank Prof. Nazareno Paolocci for his precious iconographic contribution.

Conflicts of Interest All authors declare no conflicts of interest in relation to this article.

Ethical Approval This review article does not contain any studies with human participants or animals directly performed by any of the authors.

References

Anderson B Jr, Farmer JC Jr (1978) Hyperoxic myopia. Trans Am Ophthalmol Soc 76:116–124

Baugh MA (2000) HIV: reactive oxygen species, enveloped viruses and hyperbaric oxygen. Med Hypotheses 55(3):232–238

Bergren DR, Beckman DL (1975) Hyperbaric oxygen and pulmonary surface tension. Aviat Space Environ Med 46(8):994–995

Bosco G, Yang Z, Nandi J, Wang J, Chen C, Camporesi EM (2007) Effects of hyperbaric oxygen on glucose, lactate, glycerol and antioxidant enzymes in the skeletal muscle of rats during ischemia and reperfusion. Clin Exp Pharmacol Physiol 34(1–2):70–76

Bosco G, Yang ZJ, Di Tano G, Camporesi EM, Faralli F, Savini F, Landolfi A, Doria C, Fanò G (2010) Effect of in-water versus normobaric oxygen pre-breathing on decompression-induced bubble formation and platelet activation. J Appl Physiol 108(5):1077–1083

Bosco G, Casarotto A, Nasole E, Camporesi E, Salvia R, Giovinazzo F, Zanini S, Malleo G, Di Tano A, Rubini A, Zanon V, Mangar D, Bassi C (2014) Preconditioning with hyperbaric oxygen in pancreaticoduodenectomy: a randomized double-blind pilot study. Anticancer Res 34(6):2899–2906

Bosco G, Vezzani G, Mrakic Sposta S, Rizzato A, Enten G, Abou-Samra A, Malacrida S, Quartesan S, Vezzoli A, Camporesi E (2018) Hyperbaric oxygen therapy ameliorates osteonecrosis in patients by modulating inflammation and oxidative stress. J Enzyme Inhib Med Chem 33(1):1501–1505

Camporesi EM, Bosco G (2014) Mechanisms of action of hyperbaric oxygen therapy. Undersea Hyperb Med 41 (3):247–252

ChiCTR (2020) Chinese clinical trial registry. http://www.chictr.org.cn/abouten.aspx. Accessed on 1 June 2020

Ciceri F, Beretta L, Scandroglio AM, Colombo S, Landoni G, Ruggeri A, Peccatori J, D'Angelo A, De Cobelli F, Rovere-Querini P, Tresoldi M, Dagna L, Zangrillo A (2020) Microvascular COVID-19 lung vessels obstructive thromboinflammatory syndrome (MicroCLOTS): an atypical acute respiratory distress syndrome working hypothesis. Crit Care Resusc, April 15. Online ahead of print

Clark JM, Lambertsen CJ (1971) Pulmonary oxygen toxicity: a review. Pharmacol Rev 23(2):37–133

Clark JM, Lambertsen CJ, Gelfand R, Flores ND, Pisarello JB, Rossman MD, Elias JA (1999) Effects of prolonged oxygen exposure at 1.5, 2.0; or 2.5 ATA on pulmonary function in men (predictive studies V). J Appl Physiol 1985 86(1):243–259. https://doi.org/10.1152/jappl.1999.86.1.243

Conti P, Ronconi G, Caraffa A, Gallenga CE, Ross R, Frydas I, Kritas SK (2020) Induction of pro-inflammatory cytokines (IL-1 and IL-6) and lung inflammation by coronavirus-19 (COVI-19 or SARS-CoV-2): anti-inflammatory strategies. J Biol Regul Homeost Agents 34(2):1

De Maio A, Hightower LE (2020) COVID-19, acute respiratory distress syndrome (ARDS), and hyperbaric oxygen therapy (HBOT): what is the link? Cell Stress Chaperones 18:1–4. https://doi.org/10.1007/s12192-020-01121-0. Online ahead of print

Debuc B, Smadja DM (2020) Is COVID-19 a new hematologic disease? Stem Cell Rev Rep 12:1–5. https://doi.org/10.1007/s12015-020-09987-4. Online ahead of prin

ECHM-EUBS (2020) Position statement on the use of HBOT for treatment of COVID-19 patients. http://www.eubs.org/?p=1163. Accessed on 1 May 2020

Fife CE, Piantadosi CA (1991) Oxygen toxicity. In Problems in respiratory care: clinical applications of hyperbaric oxygen. 4(2). Moon RE, Camporesi EM (Eds); JB Lippincott Co, Philadelphia; pp. 150–171

Gabrilovich DI, Musarov AL, Zmyzgova AV, Shalygina NB (1990) The use of hyperbaric oxygenation in treating viral hepatitis B and the reaction of the blood leukocytes. Ter Arkh 62(1):82–86. (Article in Russian)

Gardin C, Bosco G, Ferroni L, Quartesan S, Rizzato A, Tatullo M, Zavan B (2020) Hyperbaric oxygen therapy improves the osteogenic and vasculogenic properties of mesenchymal stem cells in the presence of inflammation in vitro. Int J Mol Sci 21(4):1452

Giamarellos-Bourboulis EJ, Netea MG, Rovina N, Akinosoglou K, Antoniadou A, Antonakos N, Damoraki G, Gkavogianni T, Adami ME, Katsaounou P, Ntaganou M, Kyriakopoulou M, Dimopoulos G, Koutsodimitropoulos I, Velissaris D, Koufargyris P, Karageorgos A, Katrini K, Lekakis V, Lupse M, Kotsaki A, Renieris G, Theodoulou D, Panou V, Koukaki E, Koulouris N, Gogos C, Koutsoukou A (2020) Complex immune dysregulation in COVID-19 patients with severe respiratory failure. Cell Host Microbe S1931–S3128(20):30236–30235

Guo D, Pan S, Wang MM, Guo Y (2020) Hyperbaric oxygen therapy may be effective to improve hypoxemia in patients with severe COVID-2019 pneumonia: two case reports. Undersea Hyperb Med 47 (2):177–183

Hadanny A, Zubari T, Tamir Adler L, Bechor Y, Fishlev G, Lang E, Polak N, Bergan J, Friedman M, Efrati S (2019) Hyperbaric oxygen therapy effects on pulmonary functions: a prospective cohort study. BMC Pulm Med 19(1):148

Halbach JL, Prieto JM, Wang AW, Hawisher D, Cauvi DM, Reyes T, Okerblom J, Ramirez-Sanchez I, Villarreal F, Patel HH, Bickler SW, Perdrize GA, De Maio A (2019) Early hyperbaric oxygen therapy improves survival in a model of severe sepsis. Am J Physiol Regul Integr Comp Physiol 317(1):R160–R168

Heyboer M, Sharma D, Santiago W, Mcculloch N (2017) Hyperbaric oxygen therapy: side effects defined and quantified. Adv Wound Care (New Rochelle) 6 (6):210–224

Hosokawa K, Yamazaki H, Nakamura T, Yoroidaka T, Imi T, Shima Y, Ohata K, Takamatsu H, Kotani T, Kondo Y, Takami A, Nakao S (2014) Successful hyperbaric oxygen therapy for refractory BK virus-associated hemorrhagic cystitis after cord blood transplantation. Transpl Infect Dis 16(5):843–846

Kim KS, Jung H, Shin IK, Choi BR, Kim DH (2015) Induction of interleukin-1 beta (IL-1β) is a critical component of lung inflammation during influenza A (H1N1) virus infection. J Med Virol 87(7):1104–1112

Kowalewski M, Fina D, Słomka A, Raffa GM, Martucci G, Lo Coco V, De Piero ME, Ranucci M, Suwalski P, Lorusso R (2020) COVID-19 and ECMO: the interplay between coagulation and inflammation-a narrative review. Crit Care 24(1):205

Li F, Fang L, Huang S, Yang Z, Nandi J, Thomas S, Chen C, Camporesi E (2011) Hyperbaric oxygenation

therapy alleviates chronic constrictive injury-induced neuropathic pain and reduces tumor necrosis factor-alpha production. Anesth Analg 113(3):626–633
Liu B, Li M, Zhou Z, Guan X, Xiang Y (2020a) Can we use interleukin-6 (IL-6) blockade for coronavirus disease 2019 (COVID-19)-induced cytokine release syndrome (CRS)? J Autoimmun 111:102452
Liu Y, Yan LM, Wan L, Xiang TX, Le A, Liu JM, Peiris M, Poon L, Zhang W (2020b) Viral dynamics in mild and severe cases of COVID-19. Lancet Infect Dis 20(6):656–657
Lodigiani C, Iapichino G, Carenzo L, Cecconi M, Ferrazzi P, Sebastian T, Kucher N, Studt JD, Sacco C, Alexia B, Sandri MT, Barco S, Humanitas COVID-19 Task Force (2020) Venous and arterial thromboembolic complications in COVID-19 patients admitted to an academic hospital in Milan, Italy. Thromb Res 191:9–14
Luo W, Yu H, Gou J, Li X, Sun Y, Li J, Liu L (2020) Clinical pathology of critical patient with novel coronavirus pneumonia (COVID-19). Preprints 2020020407
MacLaughlin KJ, Barton GP, Braun RK, Eldridge MW (2019) Effect of intermittent hyperoxia on stem cell mobilization and cytokine expression. Med Gas Res 9 (3):139–144
Marmo M, Villani R, Di Minno RM, Noschese G, Paganini M, Quartesan S, Rizzato A, Bosco G (2017) Cave canem: HBO_2 therapy efficacy on *Capnocytophaga canimorsus* infections: a case series. Undersea Hyperb Med 44(2):179–186
Mehta P, McAuley DF, Brown M, Sanchez E, Tattersall RS, Manson JJ, HLH Across Speciality Collaboration, UK (2020) COVID-19: consider cytokine storm syndromes and immunosuppression. Lancet 395 (10229):1033–1034
Milovanova TN, Bhopale VM, Sorokina EM, Moore JS, Hunt TK, Hauer-Jensen M, Velazquez OC, Thom SR (2009) Hyperbaric oxygen stimulates vasculogenic stem cell growth and differentiation in vivo. J Appl Physiol (1985) 106(2):711–728
Moon RE (2019) Hyperbaric oxygen therapy indications. 14th Edition UHMS. Best Publishing Company, North Palm Beach
Moon RE, Weaver LK (2020) Hyperbaric oxygen as a treatment for COVID-19 infection? Undersea Hyperb Med 47(2):177–179
Moon RE, Camporesi EM, Shelton DL (1987) Prediction of arterial PO_2 during hyperbaric treatment. In: Bove AA, Bachrach AJ, Greenbaum LJ Jr (eds) Underwater and hyperbaric physiology IX. Proceedings of the ninth international symposium on underwater and hyperbaric physiology. Undersea and Hyperbaric Medical Society, Bethesda, pp 1127–1131
Morabito C, Bosco G, Pilla R, Corona C, Mancinelli R, Yang Z, Camporesi EM, Fanò G, Mariggiò MA (2011) Effect of pre-breathing oxygen at different depth on oxidative status and calcium concentration in lymphocytes of scuba divers. Acta Physiol (Oxf) 202 (1):69–78
Pace GW, Leaf CD (1995) The role of oxidative stress in HIV disease. Free Radic Biol Med 19(4):523–528
Papiris SA, Tomos IP, Karakatsani A, Spathis A, Korbila I, Analitis A, Kolilekas L, Kagouridis K, Loukides S, Karakitsos P, Manali ED (2018) High levels of IL-6 and IL-8 characterize early-on idiopathic pulmonary fibrosis acute exacerbations. Cytokine 102:168–172
Pedoto A, Nandi J, Yang ZJ, Wang J, Bosco G, Oler A, Hakim TS, Camporesi EM (2003) Beneficial effect of hyperbaric oxygen pretreatment on lipopolysaccharide-induced shock in rats. Clin Exp Pharmacol Physiol 30(7):482–488
Peng Z, Wang S, Huang X, Xiao P (2012) Effect of hyperbaric oxygen therapy on patients with herpes zoster. Undersea Hyperb Med 39(6):1083–1087
Peterhans E (1997) Oxidants and antioxidants in viral diseases: disease mechanisms and metabolic regulation. J Nutr 127(5 Suppl):962S–965S
Prokof'ev VN, Mogil'nitskaia LV, Morgulis GL, Sherstneva II (1995) Biochemical composition of a surfactant and its free radical processes in hyperbaric oxygenation and in the post-hyperoxic period. Patol Fiziol Eksp Ter 3:40–43. (Article in Russian)
Rinaldi B, Cuzzocrea S, Donniacuo M, Capuano A, Di Palma D, Imperatore F, Mazzon E, Di Paola R, Sodano L, Rossi F (2011) Hyperbaric oxygen therapy reduces the toll-like receptor signaling pathway in multiple organ failures. Intensive Care Med 37 (7):1110–1119
Savva-Bordalo J, Pinho Vaz C, Sousa M, Branca R, Campilho F, Resende R, Baldaque I, Camacho O, Campos A (2012) Clinical effectiveness of hyperbaric oxygen therapy for BK-virus-associated hemorrhagic cystitis after allogeneic bone marrow transplantation. Bone Marrow Transplant 47(8):1095–1098
Schulze J, Kaiser O, Paasche G, Lamm H, Pich A, Hoffmann A, Lenarz T, Warnecke A (2017) Effect of hyperbaric oxygen on BDNF-release and neuroprotection: investigations with human mesenchymal stem cells and genetically modified NIH3T3 fibroblasts as putative cell therapeutics. PLoS One 12 (5):e0178182
Thibodeaux K, Speyrer M, Raza A, Yaakov R, Serena TE (2020) Hyperbaric oxygen therapy in preventing mechanical ventilation in COVID-19 patients: a retrospective case series. J Wound Care 29(Sup5a):S4–S8
Thom SR (2009) Oxidative stress is fundamental to hyperbaric oxygen therapy. J Appl Physiol 106(3):988–995
Thom SR (2011) Hyperbaric oxygen: its mechanisms and efficacy. Plast Reconstr Surg 127(Suppl 1):131S–141S
Thom SR, Bhopale VM, Velazquez OC, Goldstein LJ, Thom LH, Buerk DG (2006) Stem cell mobilization by hyperbaric oxygen. Am J Physiol Heart Circ Physiol 290(4):H1378–H1386

Thom SR, Bhopale VM, Mancini DJ, Milovanova TN (2008) Actin S-nitrosylation inhibits neutrophil beta2 integrin function. J Biol Chem 283(16):10822–10834

Thom SR, Bhopale VM, Yang M, Bogush M, Huang S, Milovanova TN (2011a) Neutrophil beta2 integrin inhibition by enhanced interactions of vasodilator-stimulated phosphoprotein with S-nitrosylated actin. J Biol Chem 286(37):32854–32865

Thom SR, Milovanova TN, Yang M, Bhopale VM, Sorokina EM, Uzun G, Malay DS, Troiano MA, Hardy KR, Lambert DS, Logue CJ, Margolis DJ (2011b) Vasculogenic stem cell mobilization and wound recruitment in diabetic patients: increased cell number and intracellular regulatory protein content associated with hyperbaric oxygen therapy. Wound Repair Regen 19(2):149–161

Thom SR, Bhopale VM, Milovanova TN, Yang M, Bogush M (2012) Thioredoxin reductase linked to cytoskeleton by focal adhesion kinase reverses actin S-nitrosylation and restores neutrophil β(2) integrin function. J Biol Chem 287(36):30346–30357

Thorsen E, Aanderud L, Aasen TB (1998) Effects of a standard hyperbaric oxygen treatment protocol on pulmonary function. Eur Respir J 12:1442–1445

UHMS (2020) Undersea and Hyperbaric Medicine Society. UHMS position statement: Hyperbaric Oxygen (HBO_2) for COVID-19 patients. https://www.uhms.org/images/Position-Statements/UHMS_Position_Statement_Hyperbaric_Oxygen_for_COVID-19_Patients_v13_Final_copy_edited.pdf. Accessed on 1 June 2020

US National Library of Medicine (2020) Clinical trials. https://clinicaltrials.gov/. Accessed on 1 June 2020

US Navy (2008) US Navy diving manual, 6th revision. United States: US Naval Sea Systems Command. Retrieved 2008-06-15

Webb WR, Lanius JW, Aslami A, Reynolds RC (1966) The effects of hyperbaric-oxygen tensions on pulmonary surfactant in guinea pigs and rats. JAMA 195 (4):279–280

Wong T, Wang CJ, Hsu SL, Chou WY, Lin PC, Huang CC (2008) Cocktail therapy for hip necrosis in SARS patients. Chang Gung Med J 31(6):546–553

Wu Z, McGoogan JM (2020) Characteristics of and important lessons from the coronavirus disease 2019 (COVID-19) outbreak in China: summary of a report of 72 314 cases from the Chinese Center for Disease Control and Prevention. JAMA 323(13):1239–1242. https://doi.org/10.1001/jama.2020.2648. Online ahead of print

Xu P, Zhou Q, Xu J (2020a) Mechanism of thrombocytopenia in COVID-19 patients. Ann Hematol 99 (6):1205–1208

Xu Z, Shi L, Wang Y, Zhang J, Huang L, Zhang C, Liu S, Zhao P, Liu H, Zhu L, Tai Y, Bai C, Gao T, Song J, Xia P, Dong J, Zhao J, Wang FS (2020b) Pathological findings of COVID-19 associated with acute respiratory distress syndrome. Lancet Respir Med 8 (4):420–422

Yang ZJ, Bosco G, Montante A, Ou XI, Camporesi EM (2001) Hyperbaric O2 reduces intestinal ischemia-reperfusion-induced TNF-alpha production and lung neutrophil sequestration. Eur J Appl Physiol 85 (1–2):96–103

Yang Z, Nandi J, Wang J, Bosco G, Gregory M, Chung C, Xie Y, Yang X, Camporesi EM (2006) Hyperbaric oxygenation ameliorates indomethacin-induced enteropathy in rats by modulating TNF-alpha and IL-1beta production. Dig Dis Sci 51(8):1426–1433

Adv Exp Med Biol - Clinical and Experimental Biomedicine (2021) 11: 37–54
https://doi.org/10.1007/5584_2020_556

Published online: 18 August 2020

Primary Immunodeficiencies: Diseases of Children and Adults – A Review

Aleksandra Lewandowicz-Uszyńska, Gerard Pasternak, Jerzy Świerkot, and Katarzyna Bogunia-Kubik

Abstract

Primary immunodeficiencies (PIDs) belong to a group of rare congenital diseases occurring all over the world that may be seen in both children and adults. In most cases, genetic predispositions are already known. As shown in this review, genetic abnormalities may be related to dysfunction of the immune system, which manifests itself as recurrent infections, increased risk of cancer, and autoimmune diseases. This article reviews the various forms of PIDs, including their characterization, management strategies, and complications. Novel aspects of the diagnostics and monitoring of PIDs are presented.

Keywords

Clinical symptoms · Diagnostics · Genetic defects · Primary immunodeficiency · Treatment strategy

A. Lewandowicz-Uszyńska (✉)
Third Department and Clinic of Pediatrics, Immunology and Rheumatology of Developmental Age, Wroclaw Medical University, Wroclaw, Poland

Department of Immunology and Pediatrics, The J. Gromkowski Provincial Hospital, Wroclaw, Poland
e-mail: aleksandra.lewandowicz-uszynska@umed.wroc.p

G. Pasternak
Third Department and Clinic of Pediatrics, Immunology and Rheumatology of Developmental Age, Wroclaw Medical University, Wroclaw, Poland

J. Świerkot
Department and Clinic of Rheumatology and Internal Medicine, Wroclaw Medical University, Wroclaw, Poland

K. Bogunia-Kubik
Laboratory of Clinical Immunogenetics and Pharmacogenetics, The Hirszfeld Institute of Immunology and Experimental Therapy, Polish Academy of Sciences, Wroclaw, Poland

1 Introduction

Primary immunodeficiencies (PIDs) are a group of rare congenital diseases caused by various abnormalities of immune cells, which translates to their classification (Table 1).

There are currently defined approximately 300 individual PID diseases (Tangye et al. 2020). Clinical presentation and course of these diseases depending on a specific immunological defect are presented in Table 2.

PIDs often are perceived as a problem associated solely with the developmental age, since such disorders had been commonly diagnosed and treated by pediatricians in the past. That was, at least in part, a consequence of insufficient diagnostic and therapeutic options directly affecting survival of patients until the age of 18 years. Nowadays, the situation has significantly changed, accounting for improved survival of patients with PIDs diagnosed during childhood (Table 3). The availability of high

Table 1 Basic classification of primary immunodeficiencies (PIDs), depending on the immunological abnormality, according to International Union of Immunological Societies (IUIS) (based on Bousfiha et al. 2013)

Category of disease	Diagnosis
Combined B and T cell deficiency	SCID – Severe combined immunodeficiency
	CD40 ligand deficiency (CD40L)
Prevalence of antibody deficiency	X- or AR-linked agammaglobulinemia
	CVID – Common variable immunodeficiency
	Deficiency of specific antibody
	Deficiency of IgG subclass
	Transient hypogammaglobulinemia of infancy
Other well-defined immunodeficiency syndromes	Wiskott-Aldrich syndrome
	Ataxia-telangiectasia syndrome, Nijmegen syndrome
	Hyper-IgE syndrome
	DiGeorge syndrome
Disorders of immunodeficiency with regulatory disturbance	Immunodeficiencies with hypopigmentation
	Familial hemophagocytic syndrome
	X-linked lymphoproliferative syndrome
	Autoimmune lymphoproliferative syndrome (ALPS)
Disorders of congenital phagocytosis	Severe congenital neutropenia (Kostmann syndrome)
	Cyclic neutropenia
	CGD – Chronic granulomatous disease
Deficiency of congenital immunity	Anhidrotic ectodermal dysplasia with immunodeficiency
	Deficiency of interleukin-1 receptor associated kinase 4 (IRAK-4 deficiency)
	Chronic mucocutaneous candidiasis
Auto-inflammatory disorders	Familial Mediterranean fever
	TNF receptor-associated periodic syndrome (TRAPS)
	Hyper-IgD syndrome
	Cryopyrin-associated periodic syndrome
Complement system deficiencies	Deficiencies of individual subunits of the complement system
	Hereditary angioedema (C1q-esterase inhibitor deficiency)

quality of immunodiagnostics leads to the accurate diagnosis of PID, in both children and adults.

It is currently assumed that 25–40% of all diagnoses of PIDs are made in adults (Immune Deficiency Foundation (2002)). Epidemiological studies of PIDs are rare, and particularly rare are those made in adults. The frequency of PIDs in the European Union is estimated at 1 case *per* 10,000 people (Eades-Perner et al. 2007). However, experts suggest that the real frequency of those diseases is much higher. Recent population-based studies made in South America demonstrate the estimated PID morbidity in children of 1 *per* 2,000, 1 case *per* 1,200 people, and occurring in 1 *per* 600 households (Boyle and Buckley 2007). Data estimates at the 95% confidence level suggest that the number of people with confirmed PID in the US ranges between 152,000 and 361,000. Moreover, it is estimated that 35% of PIDs are diagnoses of the common variable immunodeficiency (CVID), 26% is IgA deficiency, 13% is agammaglobulinemia, and 9% are severe combined immunodeficiency (SCID), IgG deficiency, and chronic granulomatous disease (CGD) each (Durandy et al. 2013).

The primary diagnostics facilitating the identification of PIDs is based on analytical, immunological, and genetic tests (Bonilla et al. 2015; Locke et al. 2014; Piątosa et al. 2013). Some of them may be done in majority of healthcare institutions and others in specialized laboratories only (Abraham and Aubert 2016; Routes et al.

Table 2 Clinical symptoms in primary immunodeficiencies (PIDs) depending on dysfunction of specific components of the immune system

Neutrophil count/function	T and B cells count/function	Complement subunits
Chronic periodontitis	Severe, generalized infections, resistant to antibiotics	Recurrent suppurative infections
Recurrent stomatitis	Infections with opportunistic microorganisms	Another episode of bacteremia
Recurrent skin abscess	Chronic diarrhea (resistant to treatment) caused by rotaviruses, *Campylobacter*, *Giardia lamblia*, *Cryptosporidium*, and others	Another episode of meningococcal cerebrospinal meningitis
Systemic abscesses	Chronic exfoliative erythroderma, skin ulceration	Arthritis
Staphylococcal pneumonia	Extensive skin ulceration secondary to varicella virus infection	Recurrent pneumonia and bronchitis, suppurative otitis secondary to infection with encapsulated bacteria
Osteitis	Generalized infection associated with Bacillus Calmette-Guérin (BCG) vaccine against tuberculosis; BCGitis	
Abnormal wound healing	Delayed physical development of a child	
Presence of granulomas in the respiratory, alimentary, and urinary tracts	Presence of bone lesions	
	Presence of abscesses in the central nervous system (CNS)	

Table 3 Primary immunodeficiencies (PIDs) commonly diagnosed in both children and adults

PIDs diagnosed in children	PIDs diagnosed in adults
Hypogammaglobulinemia IgG	Hypogammaglobulinemia IgG
Deficiency of IgG subclasses	Deficiency of IgG subclasses
Deficiency of specific IgG antibodies	Deficiency of specific IgG antibodies
IgA deficiency	IgA deficiency
Common variable immune deficiency (CVID)	CVID
Hyper-IgE syndrome	Hyper-IgE syndrome
Hypogammaglobulinemia IgM	
Chromosomal instability syndromes (ataxia-telangiectasia, Nijmegen breakage syndrome, bloom syndrome)	
Severe combined immunodeficiency (SCID)	
PIDs associated with severe immune defects, e.g., DiGeorge syndrome, Wiskott-Aldrich syndrome	

2014). Basic laboratory tests required for the diagnosis of PIDs are presented in Table 4.

2 Genetic Defects in Primary Immunodeficiencies (PIDs)

PIDs are a heterogeneous group of rare inherited disorders caused by a variety of monogenetic immune defects. Genetic background of numerous primary immunodeficiencies has been recently worked out, enabling a precise diagnosis, parent counselling, and prenatal tests. Mutations in over 250 different genes causing PIDs have been by far described (Al-Herz et al. 2014). Some examples are given below.

X-linked agammaglobulinemia (XLA) was described as the first congenital immunodeficiency by Bruton (1952). The disease is caused by mutations in the Bruton tyrosine kinase (BTK) encoding gene (Noordzij et al. 2002). Deficiency of BTK leads to a developmental block in the differentiation of B cells, from pro-B to pre-B. In effect, immature B cells do not leave the bone

Table 4 Basic laboratory tests useful in diagnosing primary immunodeficiencies (PIDs)

Basic analytic and bacteriology tests	Basic immunological tests for assessing
Blood cell count with smear	Immunoglobulins: IgG, IgA, IgM, IgE, IgD
Electrophoresis of proteins	IgG subclasses: IgG1, IgG2, IgG3, IgG4
C-reactive protein, procalcitonin	Specific anti-pneumococcal antibodies
Complete ionogram	Hemolytic activity of the complement
AST, ALT, GGTP,	Complement subunits: C1q-esterase, C3c and C4, C5–C9, MBL level
Cancer markers: AFP, CAE	MPO lymphocyte subpopulations:
	T – $CD3^+$, $CD4^+$, $CD8^+$,
	T native/memory CD45RA/RO,
	B – CD19, CD20, CD22,
	B mature-native – $CD19^+/CD27^-/IgD^+/IgM^+$,
	B marginal zone B cells – $CD19^+/CD27^+$
	B class-switched memory cells – $CD19^+/CD27^+/IgD^-/IgM^-$
Anti-HbsAg antibodies	NK cells – $CD3^-/CD16^+/CD56^+$
Anti-tetanus antibodies	
IgA and IgG antibodies against endomysium	Phagocytic properties of neutrophils: Phago-test
Presence of isohemagglutinins	Metabolic activity of neutrophils: Chemiluminescence, burst-test
Microalbumin	Expression of adhesion molecules: CD11a, CD18
CMV IgG and IgM – PCR method	Presence of gp91phox
EBV IgG and IgM – PCR method	Presence of double negative T cells – $CD3^+/CD^{4-}/CD8^-/TCR\alpha\beta^+$
Toxoplasma gondii IgG and IgM	Expression of CD40 and CD40L
Mycoplasma IgG, IgM	Expression of CD25 (the receptor for IL-2)
Chlamydia pneumoniae IgG, IgA, IgM	TREC level
Cultures of biological material for bacteria and fungi	Expression of MHC
	Maternal-fetal chimerism
	Presence of regulator T cells ($CD4^+/CD25^+/Foxp3^+$)
	Enzymatic analysis: ADA, PNP
	Lymphocyte function following in vivo stimulation with mitogens: PMA, ConA, PMW
	Function of lymphocytes – in vitro tests of blastic transformation with ^{3}H-thymidine

AST aspartate aminotransferase, *ALT* alanine transaminase, *GGTP* Gamma-glutamyl transferase, *MBL* mannose-binding lectin, *AFP* alpha-fetoprotein, *CAE* carcinoembryonic antigen, *MPO* myeloperoxidase, *HBsAg* hepatitis B surface antigen, *NK* nuclear killer cells, *CMV* cytomegalovirus, *EBV* Epstein-Barr virus, *TREC* T-cell receptor excision circle, *MHC* major histocompatibility complex, *ADA* adenosine deaminase, *PNP* purine nucleoside phosphorylase, *PMA* phorbol myristate acetate, *ConA* concanavalin A, *PMV* platelet-derived microvesicles

marrow, and no immunoglobulins are produced. Consequently, affected patients have a deficiency or low level of all major classes of immunoglobulins and the absence of antibody-producing plasmatic B cells in peripheral blood. They also are susceptible to various infections.

Severe combined immunodeficiency (SCID) is a group of the most severe congenital defects of immunity. The defining features usually involve a defect in the T-, B-, and NK cell systems. To date, a series of mutations within ten genes that form the SCID phenotype have been described. Inheritance of the disease is gender related (Rochman et al. 2009). This phenotype concerns 50–60% of patients who have a mutation in the gene encoding the gamma chain subunit of the interleukin-2 (IL-2) receptor (IL2RG), a component of several IL receptors such as IL-4, IL-7, IL-9, IL-15, and IL-21, and is the most common type of X-chromosome related SCID. As the chromosome X is passed on by the mother, the disease affects male offspring only. IL2RG activates an important signaling molecule, JAK3. A mutation in the JAK3 gene, located on the chromosome 19, may also lead to the development of SCID. Defective IL receptors and IL receptor pathways prevent T cells from proper development. T cells play a key role in the identification of invading pathogens and in the activation and regulation of other cells of the immune system (Piątosa et al. 2013). Other forms of SCID usually follow an autosomal recessive inheritance pattern or result from spontaneous mutations. One such form is linked to a deficiency of the enzyme adenosine deaminase (ADA), while others are caused by a variety of defects. ADA is a purine salvage enzyme expressed in all tissues. The enzyme catalyzes the irreversible deamination of 20-deoxyadenosine and adenosine to 20-deoxyinosine and inosine, respectively. ADA-SCID is an autosomal recessive disorder accounting for approximately 15–20% of all SCID cases. Mutations in the ADA gene are associated with the accumulation of adenosine and deoxyadenosine, leading to lymphopenia, expressed by decreases in T, B, and NK cells and by a wide spectrum of immune and non-immune alterations (Sauer et al. 2012). SCID is often called the "bubble boy disease". It became widely known in the 1970s when the world learned about David Vetter, a boy with X-linked SCID, who lived for 12 years in a plastic, germ-free bubble.

Wiskott-Aldrich syndrome (WAS) is a monogenic X-linked primary immunodeficiency caused by mutations in the WASP gene encoding the WAS protein (Castiello et al. 2014; Gulácsy et al. 2011). The protein is present in all blood cells and is involved in relaying signals from the surface of cells to the actin cytoskeleton – a network of fibers that make up the cell's structural framework. Signals conveyed by WASP activate the cell and trigger its movement and adhesion to other cells and tissues. In white blood cells, this signaling enables the actin cytoskeleton to establish an interaction between the immune cells and foreign invaders that they are targeted to, called "immune synapse". Mutations in the WAS gene lead to the absence of any functional WASP, which disrupts the function of the actin cytoskeleton in the developing blood cells. White blood cells that lack WASP have a decreased ability to respond to their environment and to form immune synapses, resulting in their being less able to respond to foreign invaders. Likewise, the absence of functional WASP in platelets impairs their development, leading to reduced size and premature cellular death.

Mutations in the ataxia-telangiectasia mutated (ATM) gene, situated on chromosome number 11 and responsible for the development of ataxia-telangiectasia (AT), are also referred to as the Louis-Bar syndrome (Podralska et al. 2014). The ATM gene stores instructions for making a protein that helps control cell divisions and is involved in DNA repair, as it assists cells in recognizing damaged or broken DNA strands. It coordinates DNA repair by activation of enzymes that fix DNA broken strands. Efficient repair of damaged DNA strands helps maintain the stability of the cell's genetic information. The accumulation of damaged DNA may lead to the development of malignancies. ATM gene mutations reduce or eliminate the function of the ATM protein. Without this protein, cells become unstable and die. Cerebellar neurons are particularly affected by a loss of the ATM protein. Its deficiency leads to motor problems characteristic for ataxia-telangiectasia.

PIDs are inherited in various ways such as autosomal dominant, recessive, or X-linked recessive. Autosomal dominant PIDs now account for 61 (23%) of the 260 known pathologies. Many autosomal dominant defects, such as ADA deficiency, are caused by null alleles (44/61) leading to the dominance by haploinsufficiency or negative dominance. Remaining 17 autosomal dominant defects are associated with gain-of-function mutations. Six of these 17 defective genes harbor loss-of-function mutations as well. All the known autosomal recessive PIDs are caused by alleles with some loss of function. A single X-linked recessive PID, which is WASP-related neutropenia, is caused by gain-of-function mutations (Boisson et al. 2015).

3 Modern Genetic Approaches for Molecular Diagnostics and Monitoring of Primary Immunodeficiencies (PIDs)

In addition to basic laboratory tests (Table 4), diagnosis and monitoring of PIDs can be facilitated by DNA-based analysis to detect novel mutations associated with the disease (Podralska et al. 2014) or to identify and monitor severe immunological defects, including B- or T-cell lymphopenia. DNA-based genetic approaches may be also used to study mutations and polymorphisms in other genes, which might affect the immunological response of patients with PIDs (Rezaei et al. 2009). Current diagnostic procedures for PIDs are complex and involve an array of specialized tests, including lymphocyte proliferation and cytotoxicity assays, flow cytometry, measurement of serum immunoglobulin levels, neutrophil function tests, and complement analyses (McCusker and Warrington 2011).

DNA sequencing or genotyping may be used to verify and establish a definite PID classification and to determine the optimal therapeutic strategy. A selection of candidate genes to be screened depends on patient's individual clinical and immunological characteristics. Due to a high number of PID genes, more than 200 have been described (Al-Herz et al. 2014), and relatively low prevalence of a majority of genetic defects (Ochs et al. 2014), the decision what genes or specific mutations should be assessed is not always clear. The analysis is further complicated by the fact that mutations in different genes may lead to development of similar phenotypes (locus heterogeneity), while mutations in different parts of the same gene may manifest with distinct phenotypes (allelic heterogeneity). Only is a small percentage of all PID-related genes investigated in a single laboratory. Therefore, patient's samples must be distributed to various laboratories. Recently, the development and implementation of more comprehensive diagnostic assays based on the high-throughput next generation sequencing (NGS) technologies allow for simultaneous analysis of multiple genes (Stoddard et al. 2014; Raje et al. 2014).

The usefulness of a targeted NGS approach in PID diagnostics has been recently demonstrated by Nijman et al. (2014), who used array-based and in-solution enrichment techniques combined with SOLiD sequencing for mutation screening in 170 PID genes. These methods detected point mutations and exonic deletions in 41 patients with a known genetic diagnosis. Moreover, application of NGS enabled the identification of causal mutations in 4 out of 26 patients with undiagnosed PIDs. In another study, Moens et al. (2014) have applied a selector-based target enrichment assay to detect disease-causing mutations in 179 known PID genes. The usefulness of this assay for molecular diagnosis of PID has been investigated by sequencing DNA from 33 patients, 18 of whom had at least one known causal mutation. The disease-causing mutations were identified in 60% of the investigated patients, indicating that most of the PID cases could be diagnosed using a targeted sequencing approach. The authors have proposed a stepwise approach to PID diagnostics, involving targeted resequencing followed by whole transcriptome and whole genome sequencing if causative variants cannot be found in the targeted exons. Thus, it seems that the advent of NGS technologies and target enrichment methods has led to the development of multiplex diagnostic assays. In the last decade, the number of genetically defined PIDs has increased from 106 to 260 (Al-Herz et al. 2014; Notarangelo et al. 2004), and further increase is expected to take place, along with a rapid increase in NGS application (Conley and Casanova 2014; Bolze et al. 2010).

Humoral deficiencies are the most common primary immunodeficiencies. The most severe course of the disease is observed in combined T- and B-cell immunodeficiencies. The implementation of DNA-based methods associated with detection of T-cell receptor excision circles (TRECs) and kappa-deleting recombination excision circles (KRECs) enables the identification and monitoring of severe T- and B-cell lymphopenia (Van Zelm et al. 2011). TRECs and KRECs are circularized DNA elements formed during the recombination process leading to the creation of T- and B-cell receptors (Wysoczańska 2008). Because TRECs and KRECs are unable to replicate, they persist in the cell, diluted after each cell division. Their quantity in peripheral blood may be used for the estimation of thymic and medullary output. The number of naive T lymphocytes that went through the thymic maturation process may be assessed in peripheral blood using TREC detection by quantitative PCR. Like T cells, freshly developed B cells may be assessed using KREC detection (Van Zelm et al. 2007). It has been documented that a combined TREC/KREC approach is capable of identifying patients with severe B-cell disorders, such as X-linked agammaglobulinemia, in addition to SCID patients (Borte et al. 2011a, b) and patients with DiGeorge syndrome (Froňková et al. 2014).

4 Clinical Presentation of Primary Immunodeficiencies (PIDs)

Health problems experienced by PIDs patients stem from clinical events consisting of three categories, which may occur together or singly: infections – malignancies – autoimmune diseases. Thus, affected patients require a multidisciplinary approach concerning diagnostics and therapy (Cunningham-Rundles 2011; Gompels et al. 2003).

Infections are the best known and most strongly associated with PIDs manifestation of immune dysfunction. Considering their common presence, a group of PID experts and the Jeffrey Modell Foundation have developed ten characteristic symptoms occurring in children and six symptoms occurring in adults, suggesting a diagnosis of PID (O'Sullivan and Cant 2012; Arkwright and Gennery 2011) (Table 5). These symptoms are popularized by clinical immunologists, the Polish Society for Experimental and Clinical Immunology, the Polish PID Working Group, the Polish Patient Association, operating within the framework of IMMUNOPROTECT project (Bernatowska et al. 2007), and the International Patient Organization for Primary Immunodeficiencies (IPOPI 2020).

The awareness of atypical infections or their "strange" course in PIDs accounts for improved care for that group of patients, as a shorter time to accurate diagnosis directly translates to the application of therapy. Malignancies occur significantly more commonly in the PID population, compared to the general one. Cases of leukemia and lymphoma are significantly more frequent in patients with X-linked agammaglobulinemia. IgA deficiency, particularly combined with celiac disease, is associated with a tendency for lymphoma and alimentary cancers, and CVID is clearly more commonly associated with leukemia, B-cell lymphoma, stomach cancer, colorectal carcinoma, myeloma, and thymoma (Mortaz et al. 2016; Ludvigsson et al. 2015). As a result of progress made in the therapy of infections due to effective antibiotics, immunoglobulin formulas, granulocyte colony-stimulating factors, cytokines, and transplantation of hematopoietic cells, malignancies have become the leading cause of deaths in these patients. There is a group of PIDs underlain by chromosomal instabilities, such as ataxia-telangiectasia syndrome, Nijmegen syndrome, and Bloom syndrome, where increased susceptibility of DNA strands to damages is observed under the influence of ionizing radiation (X-ray, CT) and where repair processes to eliminate those damages are defective. That situation may be a starting point for neoplastic transformation. Cases of lymphoma, leukemia, CNS malignancies, alimentary cancers, ovarian cancers, liver cancers, skin cancers, and melanoma are observed much more commonly in this group of PID patients (Chrzanowska et al. 2009; Dembowska-Baginska et al. 2009; Pietrucha et al. 2007).

Autoimmune disturbances develop also are common in PID (Azizi et al. 2016). Disturbed regulation processes, dysfunction of T cells (including recently discovered sub-populations

Table 5 Characteristic symptoms suggesting the presence of PIDs in children and adults, modified from the Jefferey Modell Foundation (https://www.info4pi.org/hq)

Pediatric symptoms	Adult symptoms
Four or more episodes of otitis a year	Four or more infections requiring an antibiotic treatment (otitis, bronchitis, sinusitis, or pneumonia) a year
Two or more episodes of severe sinusitis a year	Recurrent infections or infections requiring a long-term antibiotic treatment
Antibiotic therapy lasting for 2 months or longer, with no clear improvement	Two or more severe infections (osteitis, encephalitis, sepsis, cerebrospinal meningitis, dermatitis)
Two or more episodes of pneumonia a year	Two or more episodes of X-ray confirmed pneumonia within 3 years
Delay of normal development of a child or absence of body weight gain	Infections in atypical localizations or caused by atypical pathogens
Recurrent deep skin or systemic abscesses	Familial primary immunodeficiencies
Chronic mycosis of the oral cavity and skin in a child over 1 year of age	
Long-term use of intravenous antibiotics for infection	
Two or more severe infections or sepsis	
Family history indicating primary immunodeficiency	

of those cells: Treg and Th17), B cells, and NK cells may result in ineffective elimination of pathogens and thus account for chronic antigenic stimulation and constant activation of immunocompetent cells (Arkwright et al. 2002) (Table 6).

Chronic stimulatory activity may be responsible for regulation disorders through excessive production of cytokines, chemokines, increased expression of class II major histocompatibility complex (MHC) molecules, CD40, CD80, leading to native T-cell activation. In some cases, the phenomenon may be caused by deficit of IL-2, reduced production of tumor necrosis factor gamma (TNF-γ) abnormal proliferation of T cells, ineffective phagocytosis of immunological complexes or pathogens, and by genetic predispositions. Autoimmune disorders are mostly found in the Omen, Wiskott-Aldrich, Hyper-IgM, and DiGeorge syndromes, X-linked agammaglobulinemia, common variable immunodeficiency (CVID), and IgA deficiency (Agarwal and Mayer 2013; Gennery 2016) (Table 6).

The most common PID occurring de novo in adults is CVID, demonstrating the same morbidity in women and in men. The peak of diagnosis is at 24–34 years of age. There often is a delay of several years, usually from 3 to 8 years, from the time of onset of the first clinical symptoms to the correct diagnosis, based on the PID criteria (Dong et al. 2016). CVID is not a nosologic disease, encompassing a group of various immunity disorders. The common features present in all patients are the following:

- Deficiency of two major classes of immunoglobulins in the serum, a specific IgG below 2 x SD for age;
- Absence of isohemagglutinins and impaired post-vaccination production of specific antibodies, e.g., to tetanus anatoxin;
- Disease onset after the age of 2 years.

Deficiency of two classes of antibodies leads to persistent and recurrent infections of upper and lower respiratory tract and of the gastrointestinal tract, joints, and skin (Lewandowicz-Uszyńska et al. 2007). The course of CVID is highly individual; thus various theories regarding the etiology and familial and genetic factors that facilitate manifestation of the syndrome are proposed. Autosomal recessive, autosomal dominant, and X-linked inheritance have been demonstrated in some cases (Spickett et al. 1997). Patients often present human leukocyte antigen (HLA) determinants such as HLA-DR3, HLA-B8, and

Table 6 Most common autoimmune disorders in primary immunodeficiencies (PIDs)

PID	Autoimmune disorders
Common variable immunodeficiency (CVID)	Thrombocytopenia
	Evans syndrome
	Hemolytic or pernicious anemia
	Inflammatory bowel disease
	Neutropenia
	Rheumatoid arthritis
	Systemic lupus erythematosus
	Psoriasis
Severe combined immunodeficiencies (SCID)	Alopecia
	Dermatitis
	Thrombocytopenia
Chronic granulomatous disease (CGD)	Inflammatory bowel disease
X-linked (or Bruton's) agammaglobulinemia (XLA)	Juvenile rheumatoid arthritis
	Rheumatoid arthritis
	Dermatomyositis
Wiskott-Aldrich syndrome (WAS)	Hemolytic anemia
	Dermatitis
	Inflammatory bowel disease
	Vasculitis
Hyper-IgM syndrome	Autoimmune neutropenia
	Inflammatory bowel disease
	Rheumatoid arthritis
	Uveitis
IgA deficiency	Celiac disease
	Thrombocytopenia
	Hemolytic anemia, celiac disease
	Juvenile rheumatoid arthritis
	Chronic active hepatitis
	Systemic lupus erythematosus
	Sjogren's disease, thyroiditis
	Idiopathic Addison disease
	Crohn's disease
22q11.2 microdeletion syndrome	Autoimmune thyroiditis
	Juvenile rheumatoid arthritis
	Hemolytic anemia
Complement system deficiencies	Systemic lupus erythematosus
	Autoimmune organ-specific disorders
Immune dysregulation, polyendocrinopathy and enteropathy X-linked (IPEX)	Cytopenia
	Dermatitis
	Enteropathy
	Juvenile diabetes
Autoimmune polyendocrinopathy candidiasis ectodermal dystrophy (APECED)	Hypoparathyroidism
	Adrenal gland insufficiency (Addison disease), type 1 diabetes, Sjogren's-like syndrome

HLA-SCO1, suggesting the influence of genetic factors in the pathogenesis of this deficiency (Alper et al. 2000). There is much more common existence of IgA deficiency and strongly enhanced risk for the development of malignant tumors in families of CVID patients (Resnick et al. 2012; Castellano et al. 1992). The risk may outstandingly increase as high as 50-fold for stomach cancer and 34-fold for B-cell lymphomas (Gompels et al. 2003). Recurrent respiratory infections, such as bronchitis and pneumonia, 98% of cases, rather quickly lead to the development of chronic inflammation and fibrous lesions, resulting in bronchiectasis. Etiological factors responsible for the infections cover herpes viruses (3.6%), *Pneumocystis jiroveci* (2.8%), *Mycoplasma pneumoniae* (2.4%), and cases of sepsis caused by bacteria belonging to the genera *Pseudomonas*, *Pneumococci*, and *Haemophilus influenzae* (Bazregari et al. 2017; Martínez García et al. 2001). As it is in case of other PIDs, an important problem in CVID is increased susceptibility (approx. 20-fold or higher) to autoimmune diseases (Patuzzo et al. 2016; Świerkot and Lewandowicz-Uszyńska et al. 2007; Arkwright et al. 2002). Thus, clinical presentation may involve a spate of symptoms arising from malfunction of various organs and systems. The mean survival time in CVID patients is 55 years for women and 29 for men. Like in PIDs, the most common causes of death are malignancies, particularly non-Hodgkin's lymphomas and stomach cancer, followed by infectious complications.

The knowledge of a characteristic phenotype of PID, i.e., dysmorphic features, importantly helps in making the diagnosis. Such features are strongly expressed in chromosomal instabilities such as the Nijmegen syndrome with microcephaly, bird-like face, prominent middle part of the face, and large "button-like" eyes and the A-T syndrome with mask face, telangiectasia in the conjunctiva, ear skin, and face, and an early development of neurological disorders in the form of ataxia. Clinical presentation in both syndromes is completed by numerous "cafe au lait" spots or partial albinism (Pietrucha et al. 2018; Wolska-Kuśnierz et al. 2015). Another example of a specific phenotype is the DiGeorge syndrome manifested by facial dysmorphia consisting, inter alia, of shorter philtrum, wide-set eyes, small lower jaw, low-set ears, and a cleft lip (Boyarchuk et al. 2017).

5 Progress in Therapy of Primary Immunodeficiencies (PIDs)

Treatment of patients with PIDs, including CVID, must be interdisciplinary (Roy-Ghanta and Orange 2010). When a pathology is associated with antibody deficit, treatment is based on monthly substitution with intravenous or subcutaneous immunoglobulin formulas, with tailor-made immunoglobulin preparations, which help avoid adverse effects (Albin and Cunningham-Rundles 2014; Borte et al. 2011a, b; Lewandowicz-Uszyńska and Jankowski 2011) (Table 7).

In Poland, treatment for PIDs is reimbursed by the National Health Fund within the framework of therapeutic programs for both children and adults. Immunoglobulin formulas are administered intravenously or subcutaneously. Intravenous therapy is preferably started with a saturating dose of immunoglobulins of 500–800 mg/kg, followed by 300–500 mg/kg administered once every 30 days. For subcutaneous therapy, 100–200 mg/kg is given at weekly basis, administered with a rapid push method or by pump, usually 5 cm below the navel (Lewandowicz-Uszyńska et al. 2013; Lewandowicz-Uszyńska and Jankowski 2011). This treatment allows to maintain the serum IgG level over 450 mg/dL. In case of CVID associated with advanced inflammatory changes in the respiratory tract and bronchiectasis, application of higher intravenous IgG doses of 600 mg/kg is recommended. All bacterial infections should be intensively treated with broad-spectrum antibiotics. When infections frequently recur, despite the treatment, prophylactic administration of antibiotics or sulfonamides is recommended at half therapeutic doses. A possibility of infection with atypical bacteria should be considered and treated with appropriate antibiotics when diagnosed. Treatment of mycotic infections, e.g., with itraconazolum, fluconazolum, or posakonazolum, and parasitic infections, e.g., with metronidazolum, is essential. Patients with chronic diarrhea and absorption disorders should be treated with the elimination diet devoid of

Table 7 Treatment of primary immunodeficiencies (PIDs) patients

Prevention of infections	Treatment of auto-inflammatory and autoimmune diseases	Oncologic treatment	Prophylaxis
Intensive therapy of acute infections	Immunosuppressants: Azathioprine, leflunomide, methotrexate, mycophenolate mofetil, tacrolimus, cyclophosphamide, cyclosporine	Complete oncologic treatment	Vaccinations, particularly against encapsulated bacteria[a]
Antibiotic prophylaxis	Nonsteroidal anti-inflammatory drugs, colchicine	Radiotherapy[b]	Prohibited vaccination with viable vaccines[c]
Intravenous immunoglobulin formulas	Biological therapies: TNF inhibitors (etanercept, infliximab, adalimumab, certolizumab, golimumab); IL-1 and IL-6 antagonists (anakinra, canakinumab, rilonacept, tocilizumab); Drugs with a different mechanism of action (rituximab, abatacept, belimumab)	Transplantation of stem cells	Subcutaneous immunoglobulin formulas
Antimycotic drugs (itraconazole, fluconazole, posaconazole)	Glucocorticosteroids		Avoiding exposure to ionizing radiation if radio-susceptibility is confirmed
Anti-parasitic drugs – Metronidazole			G-CSF
			Supplementation with calcium and vit. D
			Probiotics
			Antimycotic drugs
			Counteracting malnutrition

[a]if the ability to synthesize antibodies is maintained
[b]if there are no contraindications (e.g., chromosomal instability syndromes)
[c]in absence of synthesis of antibodies and in cases of deep immunological abnormalities such as severe combined immunodeficiency (SCID), Wiskott-Aldrich syndrome (WAS), ataxia-telangiectasia syndrome, Nijmegen syndrome

gluten and disaccharide content. Probiotics restoring the normal gut microflora are recommended as well. For some PIDs, transplantation of hematopoietic cells is the only option. That mostly concerns PIDs manifesting early in life and characterized by the most severe clinical course, such as SCID, WAS, hyper-IgM, and some cases of CGD (Hassan et al. 2012; Gaspar et al. 2011).

6 Gene Therapy for Patients with Primary Immunodeficiencies (PIDs)

Immunoglobulin (antibody) replacement therapy is one of the most important and successful therapies for patients with primary immunodeficiency diseases (Shehata et al. 2010; Moore and Quinn 2008), while the best treatment for T-cell deficiency conditions is bone marrow/stem cell transplant (Antoine et al. 2003). Other treatment options, some of which are still experimental, include cytokines (Roy-Ghanta and Orange 2010), thymic transplants (Markert et al. 2010), and gene therapy (Touzot et al. 2014; Fischer et al. 2013; Mukherjee and Thrasher 2013; Pessach and Notarangelo 2011; Bogunia-Kubik and Sugisaka 2002). Patients with SCID characterized by a severe defect in both T- and B-cell systems in the natural course of the disease die during the first year of life. In 1968, the first bone marrow transplant in a patient with PID was performed. Since then, this procedure was performed in more than 1000 patients, particularly in those suffering from SCID (Gaspar et al. 2011) but also in patients with WAS, hyper-IgM, CGD, and others. The allogeneic hematopoietic stem cell (HSC) transplantation can cure many of these disorders by replacing the affected lineages with normal ones. Transplantation of HSCs or their progenitors may correct a variety of PIDs of the adaptive and innate immune systems. The success of treatment in patients with SCID largely depends on how early the therapy is implemented and how it is carried out. The cure rate reaches 95%, when transplantation is performed within the first month of life, but it drops to 75% when transplantation is delayed to the age of 3 months (Rich et al. 2001). Moreover, transplants from non-related donors are associated with increased morbidity and mortality, compared to those from an HLA-identical sibling donor. The survival for ADA-SCID patients is significantly reduced in case of a transplant from a matched non-related (66%) or haploidentical donor (43%) (Hassan et al. 2012). At present, HSC transplant from an HLA-matched sibling donor confers at least a 90% chance of cure for children suffering from SCID and about an 80% chance of cure for children affected with non-SCID PIDs (Gennery et al. 2010).

Enzyme replacement therapy with polyethylene glycol-modified adenosine deaminase (PEG-ADA) decreases a toxic ADA content and improves the immune phenotype. Nonetheless, a variable extent of immune recovery has been reported. In patients undergoing a long-term treatment, there often is a decrease in lymphocyte count, loss of regulatory T (Treg) cell function, and the development of antibodies against bovine ADA (Sauer et al. 2012; Booth and Gaspar 2009; Kohn 2008; Malacarne et al. 2005).

In the absence of a matched donor, the addition of a corrected copy of a gene into autologous HSCs is an appealing strategy. The hematopoietic system presents an optimal characteristic for the ex vivo gene therapy purpose. The gene therapy with autologous HSCs engineered with gamma retroviral vectors is a successful alternative for patients who do not have an HLA-matched donor and for whom enzyme replacement therapy is insufficient to maintain the adequate immune reconstitution (Aiuti et al. 2009). Since 2000, more than 40 patients worldwide have been enrolled into clinical trials using hematopoietic stem cell-directed gene therapy (HSC-GT), with reduced-intensity conditioning, resulting in long-term multilineage engraftment, sustained systemic detoxification, and improved immune functions.

Gene therapy is a method for treatment or prevention of genetic disorders based on delivery of repaired genes or the replacement of incorrect ones. The aim is to treat or eliminate the cause of a disease, whereas most currently used drugs just treat symptoms. In September of 1990, the US Food and Drug Administration approved the first gene therapy trial with a therapeutic goal in humans. Two children suffering from ADA-SCID, a monogenic disease leading to severe immunodeficiency, were treated with white blood cells taken from the blood of these patients and modified ex vivo to express the normal gene for adenosine deaminase. One patient exhibited a temporary response, whereas the response in the second patient was far less (Blaese et al. 1995). The gene therapy with ADA-transduced autologous stem/progenitor cells represents an alternative option for ADA-SCID patients. Since 2000, 40 patients have been treated in Italy, the UK, and the USA, with CD34+ cells transduced with the gamma-retroviral vector encoding the ADA gene, achieving a substantial clinical benefit in majority of cases. Currently, there is a wide range of target cells and diseases, including cancer, infectious, cardiovascular, monogenic (e.g., hemophilia) diseases, and rheumatoid arthritis, for which clinical studies are underway.

Clinical trials for SCID-X1 (γc deficiency) were successfully initiated in 1999. Over 60 patients with IL2RG deficiency (SCID-X1) or ADA deficiency (ADA-SCID) have received hematopoietic stem cell gene therapy in the last 15 years, using gamma retroviral vectors, resulting in immune reconstitution and clinical benefits in most cases (Hacein-Bey-Abina et al. 2010; Gaspar et al. 2004). Among the SCID forms, the T-B-SCID associated with the Artemis gene mutation is another notable candidate for gene therapy. Gene therapy involves the introduction of a "healthy" gene using the body's own stem/bone marrow cells infected with a virus that contains the correct gen. That type of treatment may be an alternative for patients who cannot find a donor of stem hematopoietic cells. Until now it has been used in SCID caused by deficiency of ADA and in SCID coupled to the X chromosome, WAS, and in CGH (Griffith et al. 2009; Rich et al. 2001; Mountain 2000; Onodera et al. 1998; Blaese et al. 1995). Several gene therapy approaches for PIDs, using initially gamma-retroviral vectors (RVs) and subsequently HIV-based lentiviral vectors (LVs), have been developed (Candotti et al. 2012). Gene therapy for PIDs seems an effective treatment, able to provide long-term clinical benefits.

Lentiviral-mediated transfer of the Wiskott-Aldrich syndrome protein (WASp) gene has been recently initiated to optimize stabilization of stem cells. It the adrenoleukodystrophy (ALD) trial (Cartier et al. 2009), 10% of all hematopoietic lineages appeared transduce-stable 3 years after therapy. In the meantime, a retrovirus-based trial for WAS has been initiated. The trial is based on the ex vivo gene transfer into CD34+ cells, following myeloablation. Preliminary results suggest that many aspects of the disease, e.g., T- and B-cell immunodeficiencies and thrombocytopenia, are corrected (Boztug et al. 2010). Although a longer follow-up is required, the results justify further clinical trials with other retroviral vectors. Self-inactivating lentiviral vector (SIN-LV) with WASp promoter is currently under trail (Merten et al. 2011).

Antisense oligonucleotides have emerged as potential gene-specific therapeutic agents able to modulate pre-mRNA splicing and to alter gene expression. Therapeutic potential of antisense oligonucleotides has been assessed in clinical trials (Bogunia-Kubik and Sugisaka 2002). The antisense oligonucleotide is a short fragment (15–20 bp) of deoxynucleotides in a sequence complementary to a portion of the targeted mRNA. The aim of the antisense strategy is to interact with gene expression by preventing the translation of proteins on mRNA. There are a few mechanisms of mRNA inactivation (Lambert et al. 2001), including (i) sterical blocking of mRNA by antisense binding and destruction of antisense-mRNA hybrids by the RnaseH enzyme, (ii) formation of triple helix between genomic double-stranded DNA and oligonucleotides, and (iii) cleavage of the target RNA by ribozymes. Many XLA-associated mutations affect splicing of Bruton's tyrosine kinase (BTK) pre-mRNA and severely impair B-cell development. The antisense, splice-correcting oligonucleotides targeting mutated BTK transcripts have been used for treating XLA. Studies show that splice-correcting oligonucleotides restore BTK function and BTK targeting in patients with XLA (Bestas et al. 2014).

7 Conclusions

PIDs are diseases of all ages; some are diagnosed in children before the age of 18 years, but others manifest as de novo diseases at later age. The practice of diagnosis serving to identify or distinguish PIDs should particularly concern patients with autoimmune diseases experiencing recurrent respiratory, gastrointestinal, or urinary tract infections. Delayed diagnosis hampers the patient's clinical condition, quality of life, and eventually survival. Most importantly, a lack of timely diagnosis hampers or even discreates the effectiveness of modern targeted molecular therapy.

Conflicts of Interest The authors declare that they have no conflict of interest in relation to this article.

Ethical Approval This review article does not contain any studies with human participants or animals performed by any of the authors.

References

Abraham RS, Aubert G (2016) Flow cytometry, a versatile tool for diagnosis and monitoring of primary immunodeficiencies. Clin Vaccine Immunol 23 (4):254–271

Agarwal S, Mayer L (2013) Diagnosis and treatment of gastrointestinal disorders in patients with primary immunodeficiency. Clin Gastroenterol Hepatol 11 (9):1050–1063

Aiuti A, Cattaneo F, Galimberti S, Benninghoff U, Cassani B, Callegaro L, Scaramuzza S, Andolfi G, Mirolo M, Brigida I, Tabucchi A, Carlucci F, Eibl M, Aker M, Slavin S, Al-Mousa H, Al Ghonaium A, Ferster A, Duppenthaler A, Notarangelo L, Wintergerst U, Buckley RH, Bregni M, Marktel S, Valsecchi MG, Rossi P, Ciceri F, Miniero R, Bordignon C, Roncarolo MG (2009) Gene therapy for immunodeficiency due to adenosine deaminase deficiency. N Engl J Med 360:447–458

Albin S, Cunningham-Rundles C (2014) An update on the use of immunoglobulin for the treatment of immunodeficiency disorders. Immunotherapy 6(10):1113–1126

Al-Herz W, Bousfiha A, Casanova JL, Chatila T, Conley ME, Cunningham-Rundles C, Etzioni A, Franco JL, Gaspar HB, Holland SM, Klein C, Nonoyama S, Ochs HD, Oksenhendler E, Picard C, Puck JM, Sullivan K, Tang ML (2014) Primary immunodeficiency diseases: an update on the classification from the international union of immunological societies expert committee for primary immunodeficiency. Front Immunol 5:162

Alper CA, Marcus-Bagley D, Awdeh Z, Kruskall MS, Eisenbarth GS, Brink SJ, Katz A, Stein R, Bing DH, Yunis EJ, Schur PH (2000) Prospective analysis suggests susceptibility genes for deficiencies of IgA and several other immunoglobulins on the [HLA-B8, SC01, DR3] conserved extended haplotype. Tissue Antigens 56:207–216

Antoine C, Muller S, Cant A, Cavazzana-Calvo M, Veys P, Vossen J, Fasth A, Heilmann C, Wulffraat N, Seger R, Blanche S, Friedrich W, Abinun M, Davies G, Bredius R, Schulz A, Landais P, Fischer A (2003) Long-term survival and transplantation of haemopoietic stem cells for immunodeficiencies: report of the European experience 1968-99. Lancet 361:553–560

Arkwright PD, Gennery AR (2011) Ten warning signs of primary immunodeficiency: a new paradigm is needed for the 21st century. Ann N Y Acad Sci 1238:7–14

Arkwright PD, Abinum M, Cant AJ (2002) Autoimmunity in human primary immunodeficiency diseases. Blood 99(8):2694–2702

Azizi G, Ahmadi M, Abolhassani H, Yazdani R, Mohammadi H, Mirshafiey A, Rezaei N, Aghamohammadi A (2016) Autoimmunity in primary antibody deficiencies. Int Arch Allergy Immunol 171 (3–4):180–193

Bazregari S, Azizi G, Tavakol M, Asgardoon MH, Kiaee F, Tavakolinia N, Valizadeh A, Abolhassani H, Aghamohammadi A (2017) Evaluation of infectious and non-infectious complications in patients with primary immunodeficiency. Cent Eur J Immunol 42 (4):336–341

Bernatowska E, Zeman K, Lewandowicz-Uszyńska A, Kurenko-Deptuch M, Pac M, Wolska-Kuśnierz B, Mikoluc B (2007) The Polish working group for primary immunodeficiency. Cent Eur J Immunol 32 (1):34–40

Bestas B, Moreno PM, Blomberg KE, Mohammad DK, Saleh AF, Sutlu T, Nordin JZ, Guterstam P, Gustafsson MO, Kharazi S, Piątosa B, Roberts TC, Behlke MA, Wood MJ, Gait MJ, Lundin KE, El Andaloussi S, Månsson R, Berglöf A, Wengel J, Smith CI (2014) Splice-correcting oligonucleotides restore BTK function in X-linked agammaglobulinemia model. J Clin Invest 124:4067–4081

Blaese RM, Culver KW, Miller AD, Carter CS, Fleisher T, Clerici M, Shearer G, Chang L, Chiang Y, Tolstoshev P, Greenblatt JJ, Rosenberg SA, Klein H, Berger M, Mullen CA, Ramsey WJ, Muul L, Morgan RA, Anderson WF (1995) T lymphocyte-directed gene therapy for ADA-SCID: initial trial results after 4 years. Science 270:475–480

Bogunia-Kubik K, Sugisaka M (2002) From molecular biology to nanotechnology and nanomedicine. Biosystems 65:123–138

Boisson B, Quartier P, Casanova JL (2015) Immunological loss-of-function due to genetic gain-of-function in humans: autosomal dominance of the third kind. Curr Opin Immunol 32:90–105

Bolze A, Byun M, McDonald D, Morgan NV, Abhyankar A, Premkumar L, Puel A, Bacon CM, Rieux-Laucat F, Pang K, Britland A, Abel L, Cant A, Maher ER, Riedl SJ, Hambleton S, Casanova JL (2010) Whole-exome-sequencing-based discovery of human FADD deficiency. Am J Hum Genet 87:873–881

Bonilla FA, Khan DA, Ballas ZK, Chinen J, Frank MM, Hsu JT, Keller M, Kobrynski LJ, Komarow HD, Mazer B, Nelson RP Jr, Orange JS, Routes JM, Shearer WT, Sorensen RU, Verbsky JW, Bernstein DI, Blessing-Moore J, Lang D, Nicklas RA, Oppenheimer J, Portnoy JM, Randolph CR, Schuller D, Spector SL, Tilles S, Wallace D (2015) Joint Task Force on Practice Parameters, representing the American Academy of Allergy, Asthma & Immunology; the American College of Allergy, Asthma & Immunology; and the Joint Council of Allergy, Asthma & Immunology. Practice parameter for the diagnosis and management of primary immunodeficiency. J Allergy Clin Immunol 136(5):1186–1205

Booth C, Gaspar HB (2009) Pegademase bovine (PEG-ADA) for the treatment of infants and children with severe combined immunodeficiency (SCID). Biologics 3:349–358

Borte S, Wang N, Oskarsdottir S, von Döbeln U, Hammarström L (2011a) Newborn screening for primary immunodeficiencies: beyond SCID and XLA. Ann N Y Acad Sci 1246:118–130

Borte M, Bernatowska E, Ochs HD, Roifman CM, Vivaglobin Study Group (2011b) Efficacy and safety of home-based subcutaneous immunoglobulin replacement therapy in paediatric patients with primary immunodeficiencies. Clin Exp Immunol 164 (3):357–364

Bousfiha AA, Jeddane L, Ailal F, Al Herz W, Conley ME, Cunningham-Rundles C, Etzioni A, Fischer A, Franco JL, Geha RS, Hammarström L, Nonoyama S, Ochs HD, Roifman CM, Seger R, Tang ML, Puck JM, Chapel H, Notarangelo LD, Casanova JL (2013) A phenotypic approach for IUIS PID classification and diagnosis: guidelines for clinicians at the bedside. J Clin Immunol 33(6):1078–1087

Boyarchuk O, Volyanska L, Dmytrash L (2017) Clinical variability of chromosome 22q11.2 deletion syndrome. Cent Eur J Immunol 42(4):412–417

Boyle JM, Buckley RH (2007) Population prevalence of diagnosed primary immunodeficiency diseases in the United States. J Clin Immunol 27:497–502

Boztug K, Schmidt M, Schwarzer A, Banerjee PP, Díez IA, Dewey RA, Böhm M, Nowrouzi A, Ball CR, Glimm H, Naundorf S, Kühlcke K, Blasczyk R, Kondratenko I, Maródi L, Orange JS, von Kalle C, Klein C (2010) Stem-cell gene therapy for the Wiskott-Aldrich syndrome. N Engl J Med 363:1918–1927

Bruton OC (1952) Agammaglobulinemia. Pediatrics 9 (6):722–728

Candotti F, Shaw KL, Muul L, Carbonaro D, Sokolic R, Choi C, Schurman SH, Garabedian E, Kesserwan C, Jagadeesh GJ, Fu PY, Gschweng E, Cooper A, Tisdale JF, Weinberg KI, Crooks GM, Kapoor N, Shah A, Abdel-Azim H, Yu XJ, Smogorzewska M, Wayne AS, Rosenblatt HM, Davis CM, Hanson C, Rishi RG, Wang X, Gjertson D, Yang OO, Balamurugan A, Bauer G, Ireland JA, Engel BC, Podsakoff GM, Hershfield MS, Blaese RM, Parkman R, Kohn DB (2012) Gene therapy for adenosine deaminase-deficient severe combined immune deficiency: clinical comparison of retroviral vectors and treatment plans. Blood 120:3635–3646

Cartier N, Hacein-Bey-Abina S, Bartholomae CC, Veres G, Schmidt M, Kutschera I, Vidaud M, Abel U, Dal-Cortivo L, Caccavelli L, Mahlaoui N, Kiermer V, Mittelstaedt D, Bellesme C, Lahlou N, Lefrère F, Blanche S, Audit M, Payen E, Leboulch P, l'Homme B, Bougnères P, Von Kalle C, Fischer A, Cavazzana-Calvo M, Aubourg P (2009) Hematopoietic stem cell gene therapy with a lentiviral vector in X-linked adrenoleukodystrophy. Science 326:818–823

Castellano G, Moreno D, Galvao O, Ballestín C, Colina F, Mollejo M, Morillas JD, Solís Herruzo JA (1992) Malignant lymphoma of jejunum with common variable hypogammaglobulinemia and diffuse nodular hyperplasia of the small intestine. A case study and literature review. J Clin Gastroenterol 15:128–135

Castiello MC, Bosticardo M, Pala F, Catucci M, Chamberlain N, van Zelm MC, Driessen GJ, Pac M, Bernatowska E, Scaramuzza S, Aiuti A, Sauer AV, Traggiai E, Meffre E, Villa A, van der Burg M (2014) Wiskott-Aldrich Syndrome protein deficiency perturbs the homeostasis of B-cell compartment in humans. J Autoimmun 50:42–50

Chrzanowska KH, Digweed M, Sperling K, Seemanova E (2009) DNA-repair deficiency and cancer: lessons from lymphoma. In: Allgayer H, Rehder H, Fulda S (eds) Hereditary tumors: from genes to clinical consequences. Wiley, Hoboken, pp 377–391

Conley ME, Casanova JL (2014) Discovery of single-gene inborn errors of immunity by next generation sequencing. Curr Opin Immunol 30C:17–23

Cunningham-Rundles C (2011) Autoimmunity in primary immune deficiency: taking lessons from our patients. Clin Exp Immunol 164(Suppl 2):6–11

Dembowska-Baginska B, Perek D, Brozyna A, Wakulinska A, Olczak-Kowalczyk D, Gladkowska-Dura M, Grajkowska W, Chrzanowska KH (2009) Non-Hodgkin lymphoma (NHL) in children with Nijmegen Breakage syndrome (NBS). Pediatr Blood Cancer 52(2):186–190

Dong J, Liang H, Wen D, Wang J (2016) Adult common variable immunodeficiency. Am J Med Sci 351 (3):239–243

Durandy A, Kracker S, Fischer A (2013) Primary antibody deficiencies. Nat Rev Immunol 13(7):519–533

Eades-Perner AM, Gathmann B, Knerr V, Guzman D, Veit D, Kindle G, Grimbacher B (2007) The European internet-based patient and research database for primary immunodeficiencies: results 2004-06. Clin Exp Immunol 147(2):306–312

Fischer A, Hacein-Bey-Abina S, Cavazzana-Calvo M (2013) Gene therapy of primary T cell immunodeficiencies. Gene 525:170–173

Froňková E, Klocperk A, Svatoň M, Nováková M, Kotrová M, Kayserová J, Kalina T, Keslová P, Votava F, Vinohradská H, Freiberger T, Mejstříková E, Trka J, Sedivá A (2014) The TREC/KREC assay for the diagnosis and monitoring of patients with DiGeorge syndrome. PLoS One 9 (1–13):e114514

Gaspar HB, Parsley KL, Howe S, King D, Gilmour KC, Sinclair J, Brouns G, Schmidt M, Von Kalle C, Barington T, Jakobsen MA, Christensen HO, Al Ghonaium A, White HN, Smith JL, Levinsky RJ, Ali RR, Kinnon C, Thrasher AJ (2004) Gene therapy of X-linked severe combined immunodeficiency by use of a pseudotyped gammaretroviral vector. Lancet 364 (9452):2181–2187

Gaspar HB, Cooray S, Gilmour KC, Parsley KL, Zhang F, Adams S, Bjorkegren E, Bayford J, Brown L, Davies EG, Veys P, Fairbanks L, Bordon V, Petropoulou T, Kinnon C, Thrasher AJ (2011) Hematopoietic stem cell gene therapy for adenosine deaminase-deficient severe combined immunodeficiency leads to long-term

immunological recovery and metabolic correction. Sci Transl Med 3(97):97ra80
Gennery AR (2016) The sting of WASP deficiency: autoimmunity exposed. Blood 127(2):173–175
Gennery AR, Slatter MA, Grandin L, Taupin P, Cant AJ, Veys P, Amrolia PJ, Gaspar HB, Davies EG, Friedrich W, Hoenig M, Notarangelo LD, Mazzolari E, Porta F, Bredius RG, Lankester AC, Wulffraat NM, Seger R, Güngör T, Fasth A, Sedlacek P, Neven B, Blanche S, Fischer A, Cavazzana-Calvo M, Landais P (2010) Transplantation of hematopoietic stem cells and long term survival for primary immunodeficiencies in Europe: entering a new century, do we do better? J Allergy Clin Immunol 126:602–610
Gompels MM, Hodges E, Rj L, Angus B, White H, Larkin A, Chapel HM, Spickett GP, Misbah SA, JL S, Associated Study Group (2003) Lymphoproliferative disease in antibody deficiency: a multi-Centre study. Clin Exp Immunol 134(2):314–320
Griffith LM, Cowan MJ, Notarangelo LD, Puck JM, Buckley RH, Candotti F, Conley ME, Fleisher TA, Gaspar HB, Kohn DB, Ochs HD, O'Reilly RJ, Rizzo JD, Roifman CM, Small TN, Shearer WT (2009) Improving cellular therapy for primary immune deficiency diseases: recognition, diagnosis, and management. J Allergy Clin Immunol 124:1152–1160
Gulácsy V, Freiberger T, Shcherbina A, Pac M, Chernyshova L, Avcin T, Kondratenko I, Kostyuchenko L, Prokofjeva T, Pasic S, Bernatowska E, Kutukculer N, Rascon J, Iagaru N, Mazza C, Tóth B, Erdos M, van der Burg M, Maródi L, J Project Study Group (2011) Genetic characteristics of eighty-seven patients with the Wiskott-Aldrich syndrome. Mol Immunol 48:788–792
Hacein-Bey-Abina S, Hauer J, Lim A, Picard C, Wang GP, Berry CC, Martinache C, Rieux-Laucat F, Latour S, Belohradsky BH, Leiva L, Sorensen R, Debré M, Casanova JL, Blanche S, Durandy A, Bushman FD, Fischer A, Cavazzana-Calvo M (2010) Efficacy of gene therapy for X-linked severe combined immunodeficiency. N Engl J Med 363:355–364
Hassan A, Booth C, Brightwell A, Allwood Z, Veys P, Rao K, Hönig M, Friedrich W, Gennery A, Slatter M, Bredius R, Finocchi A, Cancrini C, Aiuti A, Porta F, Lanfranchi A, Ridella M, Steward C, Filipovich A, Marsh R, Bordon V, Al-Muhsen S, Al-Mousa H, Alsum Z, Al-Dhekri H, Al Ghonaium A, Speckmann C, Fischer A, Mahlaoui N, Nichols KE, Grunebaum E, Al Zahrani D, Roifman CM, Boelens J, Davies EG, Cavazzana-Calvo M, Notarangelo L, Gaspar HB (2012) Outcome of haematopoietic stem cell transplantation for adenosine deaminase deficient severe combined immunodeficiency. Blood 120:3615–3624
Immune Deficiency Foundation (2002) https://web.hsd.es/redip/docs/res_his_REDIP.pdf. Primary immune deficiency diseases in America: the first national survey of patients and specialists. https://www.primaryimmune.org/publications/surveys/second_national_survey_of_patients_(2002).pdf. Accessed on 5 May 2020
IPOPI (2020) International patient organization for primary immunodeficiencies. https://www.ipopi.org. Accessed on 5 May 2020
Kohn DB (2008) Gene therapy for childhood immunological diseases. Bone Marrow Transplant 41:199–205
Lambert G, Fattal E, Couvreur P (2001) Nanoparticulate systems for the delivery of antisense oligonucleotides. Adv Drug Deliv Rev 47:99–112
Lewandowicz-Uszyńska A, Jankowski A (2011) Intravenous immunoglobulin preparations in the treatment of primary immunodeficiency. Pol Merk Lek 180:409–412
Lewandowicz-Uszyńska A, Świerkot J, Jargulińska E, Jankowski A (2007) Common variable immunodeficiency. Cent Eur J Immunol 32(1):21–26
Lewandowicz-Uszyńska A, Szaflarska A, Pietrucha B, Wójcik J, Pochylczuk R, Kuśmirek B (2013) Standards of treatment with subcutaneous immunoglobulins in primary immunodeficiencies – guidelines for doctors and nurses drawn up by three immunology centers in Poland. In: Etzioni A, Gambineri E (eds) Proceeding of the 15th meeting of the European Society Immunodeficiencies – ESID. Florence (Italy), October 3–6, 2012. Medimond S.r.l, Bologna, pp 123–125. ISBN 978-88-7587-666-1
Locke BA, Dasu T, Verbsky JW (2014) Laboratory diagnosis of primary immunodeficiencies. Clin Rev Allergy Immunol 46(2):154–168
Ludvigsson JF, Neovius M, Ye W, Hammarström L (2015) IgA deficiency and risk of cancer: a population-based matched cohort study. J Clin Immunol 35(2):182–188
Malacarne F, Benicchi T, Notarangelo LD, Mori L, Parolini S, Caimi L, Hershfield M, Notarangelo LD, Imberti L (2005) Reduced thymic output, increased spontaneous apoptosis and oligoclonal B cells in polyethylene glycol-adenosine deaminase-treated patients. Eur J Immunol 35:3376–3386
Markert ML, Devlin BH, McCarthy EA (2010) Thymus transplantation. Clin Immunol 135:236–246
Martínez García MA, de Rojas MD, Manzur NMD, Muñoz Pamplona MP, Compte Torrero L, Macián V, Perpiñá Tordera M (2001) Respiratory disorders in common variable immunodeficiency. Respir Med 95 (3):191–195
McCusker C, Warrington R (2011) Primary immunodeficiency. Allergy Asthma Clin Immunol 7:S1–S11
Merten OW, Charrier S, Laroudie N, Fauchille S, Dugué C, Jenny C, Audit M, Zanta-Boussif MA, Chautard H, Radrizzani M, Vallanti G, Naldini L, Noguiez-Hellin P, Galy A (2011) Large-scale manufacture and characterization of a lentiviral vector produced for clinical *ex vivo* gene therapy application. Hum Gene Ther 22:343–356
Moens LN, Falk-Sorqvist E, Asplund AC, Bernatowska E, Smith CI, Nilsson M (2014) Diagnostics of primary

immunodeficiency diseases: a sequencing capture approach. PLoS One 9(12):e114901

Moore ML, Quinn JM (2008) Subcutaneous immunoglobulin replacement therapy for primary antibody deficiency: advancements into the 21st century. Ann Allergy Asthma Immunol 101(2):114–121

Mortaz E, Tabarsi P, Mansouri D, Khosravi A, Garssen J, Velayati A, Adcock IM (2016) Cancers related to Immunodeficiencies: update and perspectives. Front Immunol 7:365

Mountain A (2000) Gene therapy: the first decade. Trends Biotechnol 18:119–127

Mukherjee S, Thrasher AJ (2013) Gene therapy for PIDs: progress, pitfalls and prospects. Gene 525:174–181

Nijman IJ, van Montfrans JM, Hoogstraat M, Boes ML, van de Corput L, Renner ED, van Zon P, van Lieshout S, Elferink MG, van der Burg M, Vermont CL, van der Zwaag B, Janson E, Cuppen E, Ploos van Amstel JK, van Gijn ME (2014) Targeted next-generation sequencing: a novel diagnostic tool for primary immunodeficiencies. J Allergy Clin Immunol 133:529–534

Noordzij JG, de Bruin-Versteeg S, Hartwig NG, Weemaes CM, Gerritsen EJ, Bernatowska E, van Lierde S, de Groot R, van Dongen JJ (2002) XLA patients with BTK splice-site mutations produce low levels of wild-type BTK transcripts. J Clin Immunol 22:306–318

Notarangelo L, Casanova JL, Fischer A, Puck J, Rosen F, Seger R, Geha R (2004) Primary immunodeficiency diseases: an update. J Allergy Clin Immunol 114:677–687

Ochs HD, Smith CIE, Puck JM (2014) Primary immunodeficiency diseases. A molecular and genetic approach. Oxford University Press, New York

Onodera M, Ariga T, Kawamura N, Kobayashi I, Ohtsu M, Yamada M, Tame A, Furuta H, Okano M, Matsumoto S, Kotani H, McGarrity GJ, Blaese RM, Sakiyama Y (1998) Successful peripheral T-lymphocyte-direct gene transfer for a patient with severe combined immune deficiency caused by adenosine deaminase deficiency. Blood 91:30–36

O'Sullivan MD, Cant AJ (2012) The 10 warning signs: a time for a change? Curr Opin Allergy Clin Immunol 12 (6):588–594

Patuzzo G, Barbieri A, Tinazzi E, Veneri D, Argentino G, Moretta F, Puccettio P, Lunardi C (2016) Autoimmunity and infection in common variable immunodeficiency (CVID). Autoimmun Rev 15(9):877–882

Pessach IM, Notarangelo LD (2011) Gene therapy for primary immunodeficiencies: looking ahead, toward gene correction. J Allergy Clin Immunol 127 (6):1344–1350

Piątosa B, Pac M, Siewiera K, Pietrucha B, Klaudel-Dreszler M, Heropolitańska-Pliszka E, Wolska-Kuśnierz B, Dmeńska H, Gregorek H, Sokolnicka I, Rękawek A, Tkaczyk K, Bernatowska E (2013) Common variable immune deficiency in children – clinical characteristics varies depending on defect in peripheral B cell maturation. J Clin Immunol 33:731–741

Pietrucha B, Heropolitańska-Pliszka E, Gatti RA, Bernatowska E (2007) Ataxia-telangiectasia: guidelines for diagnosis and comprehensive care. Cent Eur J Immunol 32(4):234–238

Pietrucha B, Gregorek H, Heropolitańska-Pliszka E, Bernatowska E (2018) Primary immunodeficiency with double strain break DNA (DSBs) and radiosensitvity: clinical, diagnostic and therapeutic implications. Postepy Hig Med Dosw 72:449–460. (Article in Polish)

Podralska MJ, Stembalska A, Ślęzak R, Lewandowicz-Uszyńska A, Pietrucha B, Kołtan S, Wigowska-Sowińska J, Pilch J, Mosor M, Ziółkowska-Suchanek I, Dzikiewicz-Krawczyk A, Słomski R (2014) Ten new ATM alterations in polish patients with ataxia-telangiectasia. Mol Genet Genomic Med 2:504–511

Raje N, Soden S, Swanson D, Ciaccio CE, Kingsmore SF, Dinwiddie DL (2014) Utility of next generation sequencing in clinical primary immunodeficiencies. Curr Allergy Asthma Rep 14:468

Resnick ES, Moshier EL, Godbold JH, Cunningham Rundles C (2012) Morbidity and mortality in common variable immune deficiency over 4 decades. Blood 119:1650–1657

Rezaei N, Amirzargar AA, Shakiba Y, Mahmoudi M, Moradi B, Aghamohammadi A (2009) Proinflammatory cytokine gene single nucleotide polymorphisms in common variable immunodeficiency. Clin Exp Immunol 155:21–27

Rich R, Fleisher A, Shearer T, Schroeder H, Frew A, Weyand C (2001) Clinical immunology – principles and practice. Mosby, London

Rochman Y, Spolski R, Leonard WJ (2009) New insights into the regulation of T cells by gamma(c) family cytokines. Nat Rev Immunol 9:480–490

Routes J, Abinun M, Al-Herz W, Condino-Neto A, De La Morena MT, Etzioni A, Gambineri E, Haddad E, Kobrynski L, Le Deist F, Nonoyama S, Oliveira JB, Perez E, Picard C, Rezaei N, Sleasman J, Sullivan KE, Torgerson T (2014) ICON: the early diagnosis of congenital immunodeficiencies. J Clin Immunol 34 (4):398–424

Roy-Ghanta S, Orange JS (2010) Use of cytokine therapy in primary immunodeficiency. Clin Rev Allergy Immunol 38:39–53

Sauer AV, Brigida I, Carriglio N, Aiut A (2012) Autoimmune dysregulation and purine metabolism in adenosine deaminase deficiency. Front Immunol 3:265

Shehata N, Palda V, Bowen T, Haddad E, Issekutz TB, Mazer B, Schellenberg R, Warrington R, Easton D, Anderson D, Hume H (2010) The use of immunoglobulin therapy for patients with primary immune deficiency: an evidence-based practice guideline. Transfus Med Rev 24:S28–S50

Spickett GP, Farrant J, North ME (1997) Common variable immunodeficiency: how many diseases? Immunol Today 18(7):325–328

Stoddard JL, Niemela JE, Fleisher TA, Rosenzweig SD (2014) Targeted NGS: a cost-effective approach to molecular diagnosis of PIDs. Front Immunol 5:531

Świerkot J, Lewandowicz-Uszyńska A (2007) Autoimmune disorders in the course of primary immunodeficiency. Cent Eur J Immunol 32(1):27–33

Tangye SG, Al-Herz W, Aziz BA, Chatila T, Cunningham-Rundles C, Etzioni A, Franco JL, Holland SM, Klein C, Morio T, Ochs DO, Oksenhendler E, Picard C, Puck J, Torgerson TR, Casanova JL, Sullivan KE (2020) Human inborn errors of immunity: 2019 update on the classification from the International Union of Immunological Societies Expert Committee. J Clin Immunol 40:24–64

Touzot F, Hacein-Bey-Abina S, Fischer A, Cavazzana M (2014) Gene therapy for inherited immunodeficiency. Expert Opin Biol Ther 14:789–798

Van Zelm MC, Szczepanski T, van der Burg M, van Dongen JJ (2007) Replication history of B lymphocytes reveals homeostatic proliferation and extensive antigen-induced B cell expansion. J Exp Med 204:645–655

Van Zelm MC, van der Burg M, Langerak AW, van Dongen JJM (2011) PID comes full circle: applications of V(D)J recombination excision circles in research, diagnostics and newborn screening of primary immunodeficiency disorders. Front Immunol 2:12

Wolska-Kuśnierz B, Gregorek H, Chrzanowska K, Piątosa B, Pietrucha B, Heropolitańska-Pliszka E, Pac M, Klaudel-Dreszler M, Kostyuchenko L, Pasic S, Marodi L, Belohradsky BH, Čižnár P, Shcherbina A, Kilic SS, Baumann U, Seidel MG, Gennery AR, Syczewska M, Mikołuć B, Kałwak K, Styczyński J, Pieczonka A, Drabko K, Wakulińska A, Gathmann B, Albert MH, Skarżyńska U, Bernatowska E, Inborn Errors Working Party of the Society for European Blood and Marrow Transplantation and the European Society for Immune Deficiencies (2015) Nijmegen breakage syndrome: clinical and immunological features, long-term outcome and treatment options a retrospective analysis. J Clin Immunol 35:538–549

Wysoczańska B (2008) T-cell receptor rearrangement excision circles (TRECs) as a marker of thymic function. Postepy Hig Med Dosw (Online) 62:708–724. (Article in Polish)

Adv Exp Med Biol - Clinical and Experimental Biomedicine (2021) 11: 55–62
https://doi.org/10.1007/5584_2020_562

Published online: 22 July 2020

Pneumococcal Vaccine in Adult Asthma Patients

Natalie Czaicki, Jeremy Bigaj, and Tadeusz M. Zielonka

Abstract

Streptococcus pneumoniae is the most frequent source of community-acquired bacterial pneumonia in adults. Respiratory tract infections are the foremost reasons for asthma exacerbations. The World Health Organization and the Centers for Disease Control and Prevention consider asthma a clear indication for pneumococcal vaccination. The aim of this study was to determine the extent to which adult patients with asthma in Poland adhere to a schedule of recommended pneumococcal vaccinations. In addition, the study attempted to assess the source of the patient knowledge on vaccination and the plausible determents for vaccination. The study was conducted among patients at specialist outpatient clinics in the form of an anonymous survey that contained questions about asthma, vaccines, and the knowledge and motivations. A total of 214 patients (149 females and 65 males) of the mean age of 52 ± 17 years were interviewed. A staggering 93% of patients did not receive pneumococcal vaccination, and only 24% of patients were aware of the need for this vaccine. Age, gender, and education did not affect whether patients chose to receive the vaccine. The most often quoted reason for not receiving the vaccine was lack of information, followed by lack of faith in vaccine efficacy, and the fear of adverse effects. From the standpoint of health hazard stemming from prophylaxis avoidance, it appears paramount to educate asthmatic patients on the benefits of receiving pneumococcal vaccination.

Keywords

Asthma · Health hazard · Pneumococcal vaccine · Pneumonia · Prophylaxis · Vaccination schedule

N. Czaicki
Ipswich Hospital, East Suffolk and North Essex NHS Foundation Trust, Ipswich, UK

J. Bigaj
Ysbyty Gwynedd, Betsi Cadwaladr University Health Board, Bangor, UK

T. M. Zielonka (✉)
Department of Family Medicine, Medical University of Warsaw, Warsaw, Poland
e-mail: tadeusz.zielonka@wum.edu.pl

1 Introduction

Based on the World Health Organization estimates, 300 million suffer from asthma worldwide (WHO 2015). The number of people with asthma has increased twofold over the span of the last 10 years and 255 thousand people die from asthma each year (GINA 2015). In Poland, approximately four million people (*ca* 10% population) suffer from asthma, although nearly 70% are undiagnosed and compared to the general population 5–7% of adults in Poland are asthmatics (Brożek et al. 2012). Asthma is a chronic inflammatory disease of the conducting zone of the airways, specifically the bronchi and bronchioles (GINA 2015). When triggered by an irritant, bronchial hyper-responsiveness causes reversible swelling and constriction of the bronchi, and shed epithelium creates mucous plugging. Triggers include allergens, tobacco smoke, air pollution, and viral or bacterial infections (AAAAI 2015). Respiratory tract infections are one of the foremost reasons for asthma exacerbations. *Streptococcus pneumoniae* (*S. pneumoniae*) is a Gram-positive bacterium that is the most common cause of invasive pneumonia, otitis media, septicemia, and meningitis (Skoczyńska et al. 2011). Worldwide, mortality due to *S. pneumoniae*, is approximately 1.6 million (WHO 2015). In fact, *S. pneumoniae* is the most frequent source of community-acquired bacterial pneumonia in adults in Europe. The incidence of this infection is more common among asthmatics than in those without asthma (Brożek et al. 2012). Many a study confirms the increased risk of invasive pneumococcal disease in asthmatic individuals relative to those without asthma (Kwak et al. 2015; Klemets et al. 2010; Talbot et al. 2005). A reason of increased risk of *S. pneumoniae* infection in asthmatics could be the immunological factors generally found in atopic conditions along with genetic changes in the airway infrastructure (Jung et al. 2010). Asthma can adversely affect pneumonia outcome and the related mortality rate. As such, health agencies including the Centers for Disease Control and Prevention (CDC 2015) and WHO (2015) consider asthma a clear indication for pneumococcal vaccination. Somehow differently, the Global Initiative for Asthma (GINA 2018) argues that there are still limited data that would unequivocally demonstrate the effectiveness of pneumococcal vaccine in the population of asthmatics to recommend its universal use, although admittedly such patients, particularly at advanced age, are at increased risk of contracting pneumococcal infections.

Routine immunization against *S. pneumoniae* is safe in patients with asthma, and it may mitigate a gradual decrease in respiratory function caused by repeated infectious exacerbations (Torres et al. 2015a). In Poland, pneumococcal vaccine is recommended by the health authorities for adults aged more than 50 and for patients with asthma but are not routinely reimbursed by the National Health Insurance system. The exception is free of charge availability of the 10-valent and 13-valent vaccines for high-risk patients, including but not limited to persons aged 50 or more, children aged from 2 months to 5 years, and patients diagnosed with autoimmune conditions, diabetes, HIV, chronic diseases of the heart, blood, kidneys, and lungs (Nizankowski et al. 2013). The objective of this study was to determine the extent to which adult patients with asthma in Poland adhere to a schedule of recommended pneumococcal vaccinations. In addition, the study attempted to assess the source of the patient knowledge on vaccination and the plausible determents for vaccination.

2 Methods

There were 214 patients suffering from asthma interviewed, 149 women and 65 men, mean age 52 ± 17, and range 20–91 years. Fifty-seven percent of patients were ≥ 50 years old, which put them into a high-risk group regarding pneumococcal infection. Patients were recruited from five specialist outpatient clinics in Warsaw, Poland, and were interviewed on site. The study was based on a questionnaire created specifically for patients with asthma. The diagnosis of asthma was established based on the GINA (2018). The survey was voluntary and anonymous. Data collected included age, sex, education, and causes of asthma exacerbation, whether patients had been informed to receive pneumococcal vaccination, and by whom. They were asked whether they followed physicians' recommendations regarding asthma treatment. Sixty-three percent of patients had university education (64% of women and 61% of men), 35% reported lower education, and 2% of patients failed to provide details about education. The difference in the proportion of patients with university and lower education levels amounted to 28% ($p < 0.0001$). The mean duration of asthma was 13.1 ± 14.4 years (13.7 ± 15.0 for women and 11.7 ± 12.8 for men). One half of patients had atopic asthma, confirmed by skin tests with inhalant allergens. Eighty-three percent of patients were treated by pulmonologists, 13% by allergy specialists, and 6% by family physicians. Some of the patients were being treated for asthma by more than one specialty doctor.

Data were presented as percentages of categorical variables and compared for differences with the Fisher-Irwin or chi-squared test as recommended by Campbell (2007). A p-value of <0.05 defined statistically significant differences. Data were entered into a database using Microsoft Excel v2011.

3 Results

Sixty-three percent of patients reported at least one incident of exacerbation of asthma within the last 2 years, and 59% of patients reported an infection as a cause of exacerbation. Other causes were exercise listed by 42% of patients and allergens listed by 37%. Further causes had an environmental background such as bad weather (12%) and pollution (12%). Some patients were unsure as to what aggravated their asthma. In fact, 81% of patients stated they had at least one respiratory tract infection in the past year.

Referring to the pneumococcal vaccine, only were 24% of patients aware they could qualify to receive it. Despite that a staggering 93% of the patients surveyed did not receive the vaccine. There was a notable gender difference in the vaccine awareness, which was perceived by 29% females and 18% males ($p = 0.04$) (Fig. 1). Despite the difference, a similar but small 4% proportion of both female and male patients was vaccinated. Patient age did not affect significantly the level of vaccine awareness as 24% of patients aged over 50 and 23% under 50 reported being aware. Among patients who received the vaccine, 3% were over 50, and 5% were under 50 (Table 1).

Awareness of the recommendation to become vaccinated against pneumococcus was reported by 13% of patients having university education and by 5% of those with lower levels of education. However, there was no significant difference in the level of education between the patients who received and not received the vaccine. Three percent of vaccinated patients had university education, and 3.9% had lower education. Age, gender, and education did not affect the patient's decision to receive the vaccine.

In 80% of cases, patients with asthma were treated by respiratory physicians (Fig. 1a). However, only did about 70% of those physicians recommend pneumococcal vaccination (Fig. 2b). The patients' decision to become vaccinated was also based on recommendations given by family physicians (27%) and allergy physicians (9%) (Fig. 2b). Nine percent of patients became vaccinated on their own, usually getting information from the Internet. Overwhelmingly, the most common reason for not complying with vaccine

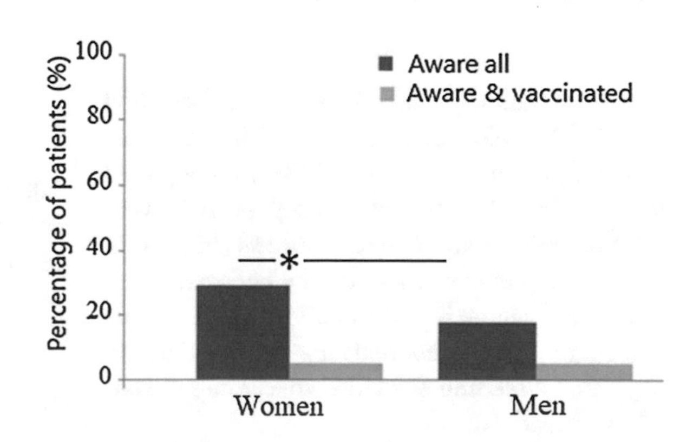

Fig. 1 Patients with asthma aware of the recommendation for pneumococcal vaccination by gender versus the proportion of those truly vaccinated; *$p = 0.04$

Table 1 Awareness of availability of pneumococcal vaccine in patients with asthma by gender and age

	Women (n = 149)		Men (n = 65)		Mean
Awareness of vaccine Irrespective of becoming vaccinated	≥50 years old	28%	≥50 years old	20%	24%
	<50 years old	30%	<50 years old	15%	
	Mean	29%	Mean	18%	
Awareness of vaccine, becoming vaccinated	≥50 years old	3%	≥50 years old	3%	7%
	<50 years old	5%	<50 years old	4%	
	Mean	4%	Mean	4%	
Awareness of vaccine, not becoming vaccinated	≥50 years old	19%	≥50 years old	17%	17%
	<50 years old	25%	<50 years old	12%	
	Mean	22%	Mean	14%	
Adherence to physician's recommendations	≥50 years old	93%	≥50 years old	93%	89%
	<50 years old	82%	<50 years old	93%	
	Mean	88%	Mean	91%	

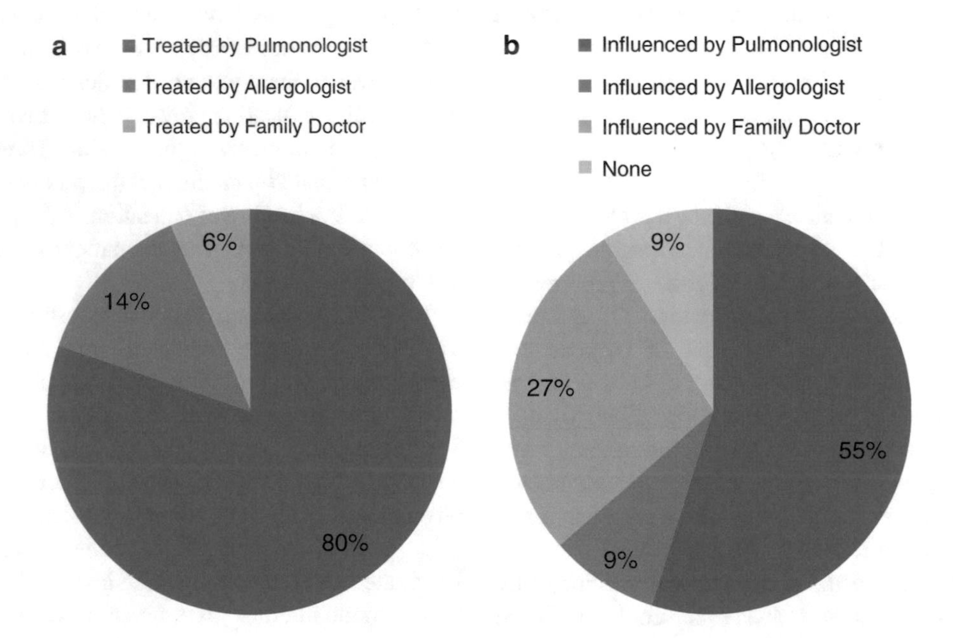

Fig. 2 Specialty physicians treating patient with asthma (**a**); physician who influenced patients' decision to become vaccinated against *S. pneumoniae* (**b**)

recommendations was lack of information reported by 76% of patients, which was overwhelmingly more than the 14% who reported lack of faith in vaccine efficacy or 10% who were apprehensive of adverse effects (Fig. 3). In fact, some patients reported they believed pneumococcal vaccine is just for children. Eighty-nine percent of patients reported they usually adhere to physician's recommendations concerning the use of prescribed drugs.

4 Discussion

Talbot et al. (2005) have demonstrated that asthma is an independent risk factor for invasive pneumococcal disease and the risk for developing a pneumococcal disease in asthmatics is at least twice that present in healthy subjects. Nonetheless, there have been very few studies that address the pneumococcal vaccination rate in asthmatic

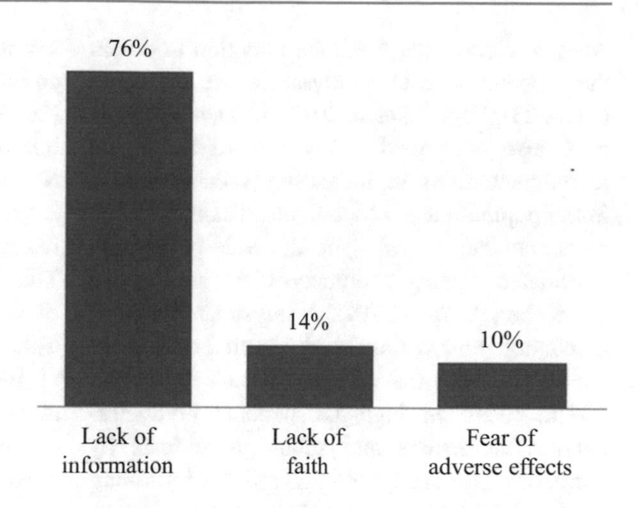

Fig. 3 Reasons for not receiving pneumococcal vaccination declared by patients with asthma

adults. Sheikh et al. (2002) have attempted a meta-analysis on the subject, retrieving just three papers with only one hardly satisfying inclusion criteria. Thus, no data could be aggregated in the analysis. By the end of 2015, 129 countries have initiated vaccination against *S. pneumoniae*, with global coverage of about 37% (WHO 2015). The findings of the study show a diminutive 7% proportion of asthmatics were vaccinated against pneumococcal infection in Poland. Meanwhile, the US projections for vaccination in 2020 reach 60% asthmatics (US DHHS 2020). Adults with work-related asthma are more likely to receive pneumococcal vaccination than those with other types of asthma −54% vs. 35%, respectively (Dodd and Mazurek 2017). An Australian study has reported that out of 2203 high-risk adult patients, i.e., those with asthma, diabetes, or cardiovascular disorders, 31% are immunized against pneumococcus. That study also shows that immunization coverage increases with age as the number of immunized patients aged 65+, who incidentally receive vaccine free of charge, overrides that present in the general population. The authors conclude that providing a free vaccination to all high-risk people could improve immunization coverage in younger patients as well (Dower et al. 2011). Interestingly, asthmatic patients in advanced age appear not to get vaccinated more often than middle-aged asthmatics. In fact, advanced age is an indication for pneumococcal vaccination, regardless of chronic pathologies such as asthma (CDC 2015). This seems to demonstrate insufficiency of adequate education on the part of both specialists and general practitioners alike, who ought to recommend vaccination for the elderly. Although an increased number of women are aware of the need to get vaccinated, as shown in the present study, this does not translate into a higher vaccination rate when compared to men. Likewise, a higher level of education, often going in hand with being economically better off, does not translate into a higher vaccination rate. In the present study, patients declared that the main reason for vaccination was a physician's recommendation. Conversely, the main reason for getting not vaccinated was a lack of such recommendation. This shows a paramount role of caregivers in persuading patients to pursue prophylactic measures. In Poland, extensive staff shortages result in physicians having an exceptionally limited amount of time with the patient. That explains, at least partly, a low proportion of asthmatic patients who get vaccinated against pneumococci. The results also indicate a need to find a way to motivate physicians to advise their patients the appropriate vaccines. This is also confirmed by a Japanese study that indicates a family physician's recommendation has been positively correlated with the intention to obtain the pneumococcal polysaccharide vaccine (PPSV23) (Higuchi et al. 2018). Canadian studies have also confirmed a key role of healthcare recommendations in increasing vaccine uptake in the population of older adults. Thanks to such recommendations 58% of the elderly became vaccinated against pneumococci in a study by Schneeberg et al. (2014). A physician's recommendation and patient's education go hand in hand to enhance the vaccination rate. Physicians should focus on high-risk patients in doctor-patient interactions and should ease fears by correcting misconceptions and reliably informing on the risks and benefits. Electronic reminders, prompting physicians to offer vaccines, also appear helpful (Chang 2015).

In Europe, national immunization programs and recommendations, apart from infant immunization standards, vary from country to country. Recommendations are usually based on risk or age. For instance, France, Sweden, and the UK recommend the pneumococcal vaccine for asthmatic and immunocompromised patients and for smokers. Furthermore, Sweden and the UK have age-based recommendations for individuals aged 65+ (Torres et al. 2015b). German and Italian studies have found that adult pneumococcal vaccination is beneficial from the stand point of health and cost-effectiveness (Boccalini et al. 2013; Kuhlmann et al. 2012). A Polish study, which assessed the coverage of pneumococcal vaccination across Europe, has concluded that compulsory anti-pneumococcal inoculation of children in Slovakia results in the highest percentage of immunization. In other European countries, which do not exercise a compulsory inoculation, such as France, Sweden, or the UK, vaccination rate is lower than that in Slovakia (89%, 60%, and 90%, respectively) (Nizankowski et al. 2013). In Poland, pneumococcal vaccination has not been compulsory; thus, a small proportion of children is vaccinated. A delayed effect of that is increased risk of contracting *S. pneumoniae* by adults, as children are often a source of infection that spreads to the adult population: an infection that may run a severe course in the elderly and patients with co-morbidities. A study by Patrzałek et al. (2016) has demonstrated that the experimental introduction of compulsory vaccination against *S. pneumoniae* for all children younger than 2 years in one town caused a 67% decrease of death rates from pneumonia in adults aged 65+. Thus, in the absence of compulsory vaccination of children against pneumococci in Poland, asthmatics, especially those in advanced age, are at increased risk of deaths from contracting *S. pneumoniae*.

The present study demonstrates that the anti-pneumococcal vaccination rate among patients diagnosed with asthma is dismally low (7%) in Poland. Patients received a single dose of a 23-valent pneumococcal polysaccharide vaccine. Recently, a 13-valent pneumococcal conjugate vaccine has been introduced for adults, which may likely become standard in the future. A substantially greater proportion of asthmatics gets vaccinated against the flu. This proportion ranges from 12% (Lu and Nuorti 2012) to 43% in the USA (Mazurek et al. 2014) and is 35% in Spain (Santos–Sancho et al. 2013) and 47% in Australia (Dower et al. 2011). Incidentally, in the asthmatic patients investigated in the present study, 20% regularly received the flu vaccine and 43% reported receiving the vaccine at least once in their life.

In conclusion, this study shows that the awareness of pneumococcal vaccine recommendation is low in Polish patients with asthma, due mostly to a lack of reliable information. Unsurprisingly, the patient compliance is very low as well. We found that the most common source of information comes from physicians. It is paramount to reliably educate patients about the indications and benefits of receiving pneumococcal vaccination, especially in relation to their asthma. Information given by physicians is usually seriously considered and may substantially enhance the vaccination coverage rate.

Acknowledgments The authors thank Drs. Andrzej Chcialowski, Piotr Dabrowiecki, and Jan Oleksy for their help and efforts in collecting data from patients for this study.

Conflicts of Interest The authors declare no conflict of interest in relation to this article.

Ethical Approval All procedures performed in studies involving human participants were in accordance with the ethical standards of the institutional and/or national research committee, and with the 1964 Helsinki declaration and its later amendments or comparable ethical standards, the Ethics Committee of Warsaw Medical University in Warsaw, Poland, approved the study protocol.

Informed Consent Informed consent was obtained from all individual participants included in the study.

References

AAAAI (2015) American Academy of Allergy, Asthma and Immunology. Asthma. https://www.aaaai.org/conditions–and–treatments/asthma.aspx. Accessed on 28 Oct 2018

Boccalini S, Bechini A, Levi M, Tiscione E, Gasparini R, Bonanni P (2013) Cost–effectiveness of new adult pneumococcal vaccination strategies in Italy. Hum Vaccin Immunother 9(3):699–706

Brożek GM, Nowak M, Pierzchała W, Zejda JE (2012) Profile of adults suffering from asthma in Poland – results of PulmoScreen study. Pneumonol Alergol Pol 80:402–411

Campbell I (2007) Chi-squared and Fisher-Irwin tests of two-by-two tables with small sample recommendations. Stat Med 26:3661–3675

CDC (2015) Centers for Disease Control and Prevention. Pneumococcal vaccination. https://www.cdc.gov/vaccines/vpd–vac/pneumo/default.htm?s_cid=cs_797. Accessed on 28 Oct 2018

Chang QCT (2015) An audit on pneumococcal vaccination rates among adult patients with asthma and chronic obstructive pulmonary disease (COPD) in a local polyclinic. Antimicrob Resist Infect Control 4 (Suppl 1):P108

Dodd KE, Mazurek JM (2017) Pneumococcal vaccination among adults with work–related asthma. Am J Prev Med 53(6):799–809

Dower J, Donald M, Begum N, Valck S, Ozolins I (2011) Patterns and determinants of influenza and pneumococcal immunisation among adults with chronic disease living in Queensland, Australia. Vaccine 29 (16):3031–3037

GINA (2015) Global Initiative for Asthma. Global strategy for asthma management and prevention. Updated 2015. https://www.ginasthma.org/local/uploads/files/GINA_Report_2015.pdf. Accessed on 20 May 2018

GINA (2018) Global Initiative for Asthma. Global strategy for asthma management and prevention: online appendix 2015. http://www.ginasthma.org/local/uploads/files/GINA_Appendix_2018_May19.pdf. Accessed on 28 Oct 2019

Higuchi M, Narumoto K, Goto T, Inoue M (2018) Correlation between family physician's direct advice and pneumococcal vaccination intention and behavior among the elderly in Japan: a cross-sectional study. BMC Fam Pract 19(1):153

Jung J, Kita H, Yawn B, Boyce T, Yoo K, McGree M, Weaver A, Wollan P, Jacobson R, Juhn Y (2010) Increased risk of serious pneumococcal disease in patients with atopic conditions other than asthma. J Allergy Clin Immunol 125:217–221

Klemets P, Lyytikäinen O, Ruutu P, Ollgren J, Kaijalainen T, Leinonen M, Nuorti JP (2010) Risk of invasive pneumococcal infections among working age adults with asthma. Thorax 65:698–702

Kuhlmann A, Theidel U, Pletz MW, von der Schulenburg JM (2012) Potential cost–effectiveness and benefit–cost ratios of adult pneumococcal vaccination in Germany. Health Econ Rev 2(1):4

Kwak BO, Choung JT, Park YM (2015) The association between asthma and invasive pneumococcal disease: a nationwide study in Korea. J Korean Med Sci 30:60–66

Lu PJ, Nuorti JP (2012) Uptake of pneumococcal polysaccharide vaccination among working–age adults with underlying medical conditions, United States, 2009. Am J Epidemiol 175(8):827–837

Mazurek JM, White GE, Moorman JE, Storey E (2014) Influenza vaccination among persons with work–related asthma. Am J Prev Med 47(2):203–211

Nizankowski R, Koperny M, Kargul A, Seweryn M (2013) Availability of pneumococcal vaccination programmes in Europe. J Health Policy Outcomes Res 1:38–49

Patrzałek M, Kotowska M, Goryński P, Albrecht P (2016) Indirect effects of a 7 year PCV7/PCV13 mass vaccination program in children on the incidence of pneumonia among adults: a comparative study based on two Polish cities. Curr Med Res Opin 32(3):397–403

Santos–Sancho JM, Lopez-de Andres A, Jimenez-Trujillo-I, Hernandez-Barrera V, Carrasco-Garrido P, Astasio-Arbiza P, Jimenez-Garcia R (2013) Adherence and factors associated with influenza vaccination among subject with asthma in Spain. Infection 41(2):465–471

Schneeberg A, Bettinger JA, McNeil S, Ward BJ, Dionne M, Cooper C, Coleman B, Loeb M, Rubinstein E, McElhaney J, Scheifele DW, Halperin SA (2014) Knowledge, attitudes, beliefs and behaviours of older adults about pneumococcal immunization, a public Health Agency of Canada/Canadian Institutes of Health Research Influenza Research Network (PCIRN) Investigation. BMC Public Health 14:442

Sheikh A, Alves B, Dhami S (2002) Pneumococcal vaccine for asthma. Cochrane Database Syst Rev 1: CD002165

Skoczyńska A, Sadowy E, Bojarska K, Strzelecki J, Kuch A, Gołębiewska A, Waśko I, Foryś M, van der Linden M, Hryniewicz W (2011) The current status of invasive pneumococcal disease in Poland. Vaccine 29:2199–2205

Talbot TR, Hartert TV, Mitchel E, Halasa NB, Arbogast PG, Poehling KA, Schaffner W, Craig AS, Griffin MR (2005) Asthma as a risk factor for invasive pneumococcal disease. N Engl J Med 352(20):2082–2090

Torres A, Blasi F, Dartois N, Akova M (2015a) Which individuals are at increased risk of pneumococcal disease and why? Impact of COPD, asthma, smoking, diabetes, and/or chronic heart disease on community–acquired pneumonia and invasive pneumococcal disease. Thorax 70:984–989

Torres A, Bonanni P, Hryniewicz W, Moutschen M, Reinert R, Welte T (2015b) Pneumococcal vaccination: what have we learnt so far and what can we expect in the future? Eur J Clin Microbiol Infect Dis 34:19–31

US DHHS (2020) Topics & objectives. Immunization and infectious diseases. https://www.healthypeople.gov/2020/topics–objectives/topic/immunization–and–infectious–diseases/objectives. Published September 22, 2016. Accessed on 23 Sept 2019

WHO (2015) Pneumococcal disease. https://www.who.int/ith/diseases/pneumococcal/en/. Accessed on 21 Nov 2018

Adv Exp Med Biol - Clinical and Experimental Biomedicine (2021) 11: 63–70
https://doi.org/10.1007/5584_2020_541

Published online: 10 June 2020

Immunoglobulin G Deficiency in Children with Recurrent Respiratory Infections with and Without History of Allergy

Aleksandra Lewandowicz-Uszyńska, Gerard Pasternak, and Katarzyna Pentoś

Abstract

Recurrent respiratory tract infections (RTI) are one of the most common diseases in childhood. Frequent infections adversely affect the development of a child and may lead to suspicion of immunodeficiency. An additional allergy component is thought conducive to infection occurrence. In this study, we retrospectively assessed medical records of 524 children hospitalized with RTI. Patients were divided into two groups: RTI-alone ($n = 394$) and RTI with a history of allergy ($n = 130$). Overall, we found that a great majority of children with RTI had the immunoglobulin G within the normal limit, irrespective of allergy. A variable IgG deficiency, most often affecting IgG1, IgG3, and IgG4 subclass, was present in less than one-third of children. Proportions of specific IgG subclass deficiency, varying from about 10% to 40%, were similar in both RTI-alone and RTI-allergy groups. The only significant effect was a modestly smaller proportion of children with IgG4 deficiency in the RTI-allergy group when compared with the RTI-alone group. We also found that IgG deficiencies were age-dependent as their number significantly increased with children's age, irrespective of allergy. The results demonstrate a lack of distinct abnormalities in the immunoglobulin G profile which would be characteristic to a clinical history of allergy accompanying recurrent RTI in children. Thus, we conclude that the assessment of IgGs could hardly be of help in the differential diagnostics of the allergic background of RTI.

Keywords

Allergy · Children · Humoral disorders · IgG subclasses · Immune deficiency · Immunity · Respiratory infections

1 Introduction

Studies on the concentrations level of IgG subclasses are important in the assessment of humoral immunity. This is due, inter alia, to the biological properties of IgG that have an influence on other elements of the immune system. IgG3, followed by IgG1, most strongly induces complement by the ability of CH2 domains of IgG to bind to both IgG-Fc receptors (FcγR) and

A. Lewandowicz-Uszyńska and G. Pasternak (✉)
Third Department and Clinic of Pediatrics, Immunology and Rheumatology of Developmental Age, Wroclaw Medical University, Wroclaw, Poland

Department of Immunology and Pediatrics, J. Gromkowski Regional Specialist Hospital, Wroclaw, Poland
e-mail: gerard.pasternak@umed.wroc.pl

K. Pentoś
Institute of Agricultural Engineering, Wroclaw University of Environmental and Life Science, Wroclaw, Poland

C1q. That is the way in which the IgG proteins contribute to the efficiency of phagocytosis and to cell activation by antibody-dependent cytotoxicity (ADCC), thereby increasing the bactericidal potential of the body (Michaelsen et al. 2009). The subclasses IgG2 and IgG1, and to a smaller extent IgG3 and IgG4, are involved in the response to polysaccharide antigens (Vidarsson et al. 2014). The IgG subclasses also have the ability to bind bacterial proteins such as staphylococcal protein A (IgG subclasses 1, 2, and 4) and streptococcal G proteins (all 4 IgG subclasses) (Falconer et al. 1993). Likewise, the ability to penetrate the placenta concerns all subclasses, which significantly protects the fetus and newborn against infection. The principal goal of this study was to define the content of IgG subclasses in children with recurrent respiratory tract infections (RTI) and a negative or positive history of allergy.

2 Methods

This is a retrospective study in which 524 medical files were reviewed under the angle of changes in IgG immunoglobulin subclasses in children hospitalized due to recurrent RTI. Patients were divided into two groups: RTI-alone ($n = 394$) and RTI accompanied by symptoms of allergy ($n = 130$). The RTI-alone group included children who had at least eight infections of the upper or lower respiratory tract within the last 12 months, in particular bronchitis and pneumonia. The RTI-allergy group consisted of children with RTI and clinical allergy symptoms such as polinosis, bronchial asthma, or atopic dermatitis, who were cared for by an allergy specialist. The children included in the study were aged from 0 to 18 years. The RTI group consisted of 152 (38%) females and 242 (62%) males, and the RTI-allergy group consisted of 44 (34%) females and 86 (66%) males. Due to a large age span and the resulting differences in the concentrations of IgG and IgG subclasses, the following age subgroups of children were distinguished:

RTI-alone group

- 266 (67%) under 6 years
- 95 (24%) between 6 and 12 years
- 33 (9%) between 12 and 18 years

RTI-allergy group

- 68 (52%) under 6 years of age
- 42 (33%) between 6 and 12 years
- 20 (15%) between 12 and 18 years

The results of serum immunoglobulins were evaluated taking into consideration the reference values for gender across age groups investigated in the pediatric population. The level of total IgG (mg/dL) was assessed with an immunoturbidimetric analyzer (Architect c-System; Abbott Laboratories, Lake Bluff, IL). The IgG1, IgG2, IgG3, and IgG4 subclasses were assessed with a nephelometric analyzer (BN ProSpec System; Siemens Healthcare GmbH, Erlangen, Germany).

Data were shown as counts and percentages of children. The distribution of data was evaluated using the Shapiro-Wilk test. Statistical differences were assessed using the Mann-Whitney U test and a Chi^2 test. A p-value <0.05 defined statistically significant changes. The analysis was performed using a commercial Statistica v13 package (StatSoft; Tulsa, OK).

3 Results

3.1 IgG Deficiencies in Children with Respiratory Tract Infection with and Without a History of Allergy

We found variable deficiencies in IgG subclasses in some of the children in both RTI-alone and RTI-allergy. The percentage distribution of children with IgG subclass deficiencies is illustrated in Fig. 1. There were about 30% of children with IgG1 deficit, 6% with IgG2 deficit, 20% with IgG3, 14–20% with IgG4 deficit, and 13% with

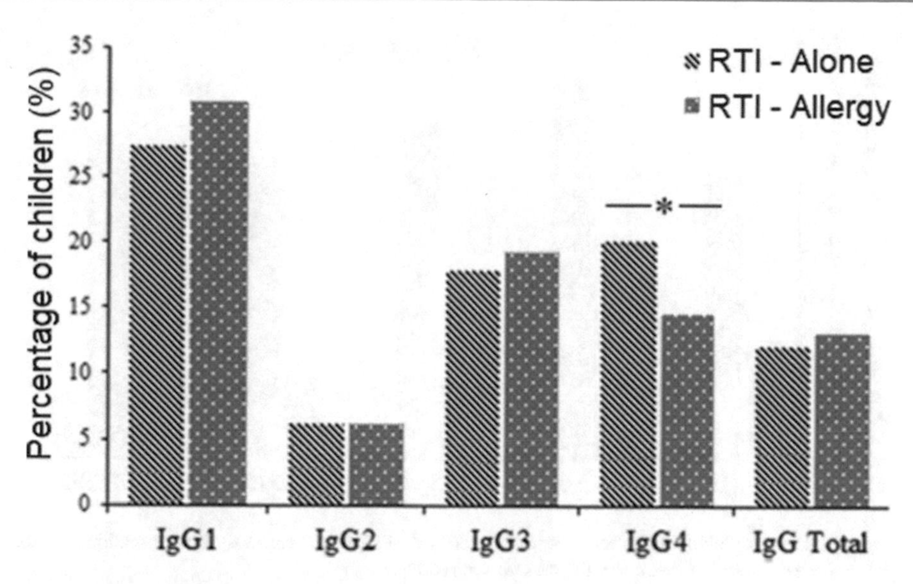

Fig. 1 Distribution of immunoglobulin G deficiencies in children with recurrent respiratory tract infections without (RTI-alone) and with a history of allergy (RTI – Allergy); $*p < 0.05$

total IgG deficit. However, the percentage of children deficient in a specific IgG subclass was similar in either group, except the IgG4 subclass where it amounted to 14% of children in RTI-allergy and 20% of children in RTI-alone ($p < 0.05$).

Despite the IgG deficiencies above outlined, we found that the approximately 65–90% majority of children in either group of recurrent RTI had the IgG subclass levels within the normal limit, with a fraction of them having IgG enhanced (Figs. 2 and 3). The IgG2 was a subclass placed within the normal limit in about 90% of children in both groups. In contrast, the IgG1 was a subclass most clearly decreased, followed by IgG3 and IgG4, irrespective of the presence of allergy. On the other hand, IgG4 was a subclass most often increased above the normal limit in both groups of children, reaching about 15% of RTI-allergy children (Fig. 3). However, there were no significant differences in the distribution profile of children with IgG changes.

Some of the children had a concomitant deficiency in two IgG subclasses. The most frequent double deficiency concerned the IgG1 + IgG3 combination, which concerned 9% and 11% of children, followed by IgG1 + IgG4 deficiency, which concerned 6% and 5% of children, in the RTI-alone and RTI-allergy group, respectively (Table 1). About 1% of children in either group were deficient in all IgG subclasses. On the percentage basis, there were no significant differences between the two groups in the proportion of children with specific pairs of double IgG subclass deficiency. In the absolute terms, however, the numbers of children with IgG1 + IgG3 and IgG1 + IgG4 in the RTI-alone and IgG1 + IgG3 deficiencies in the RTI-allergy group were significantly greater than those in the remaining double-subclass pairs ($p < 0.05$).

We further separated out 54 children from the RTI-allergy group, suffering from asthma. This subgroup was no different in terms of total IgG and IgG-specific subclass deficiencies when compared to the remaining children in either group investigated (data not shown).

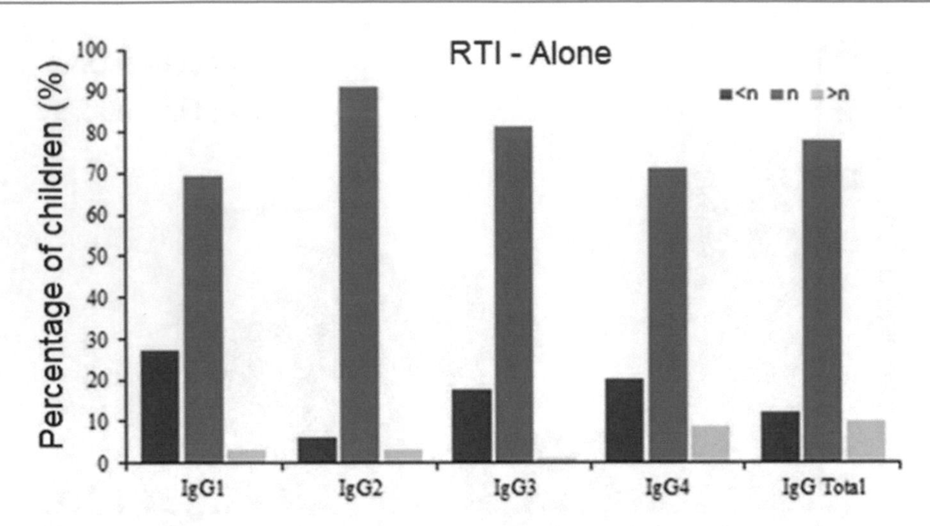

Fig. 2 Changes in immunoglobulin G subclasses in children with recurrent respiratory tract infections without a history of allergy (RTI -alone); *n,* normal level; *<n,* decreased; *>n,* increased

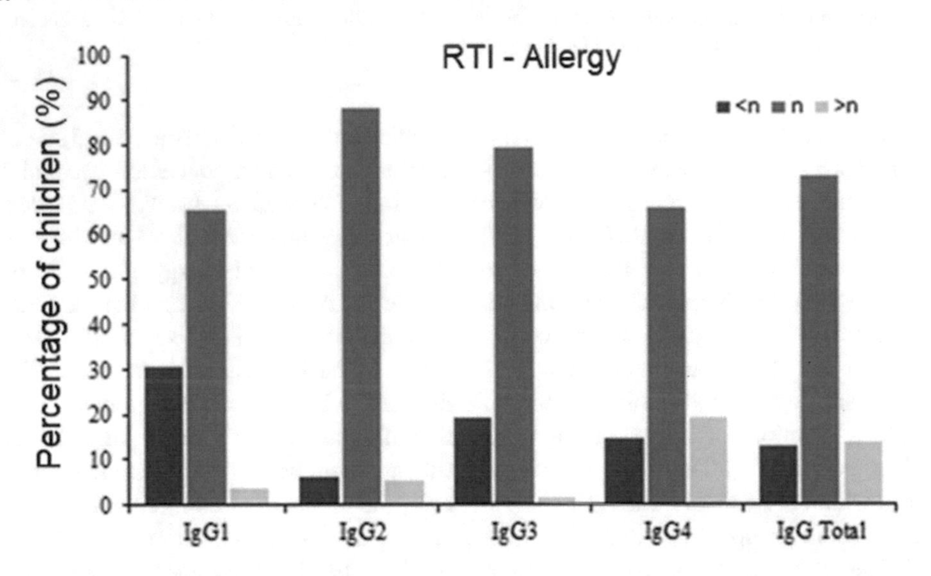

Fig. 3 Changes in immunoglobulin G subclasses in children with recurrent respiratory tract infections with a history of allergy (RTI-allergy); *n* normal level; *<n,* decreased; *>n,* increased

Table 1 Double immunoglobulin G deficiency in children with recurrent respiratory tract infections without (RTI-alone) and with a history of allergy (RTI-allergy)

	RTI-alone	RTI-allergy
IgG subclasses	*n (%)*	*n (%)*
IgG1 + IgG2	11 (3)	5 (4)
IgG1 + IgG3	36 (9)	14 (11)
IgG1 + IgG4	24 (6)	6 (5)
IgG2 + IgG3	10 (3)	3 (2)
IgG2 + IgG4	12 (3)	5 (4)
IgG3 + IgG4	14 (4)	3 (2)

3.2 IgG Deficiencies in Children with Respiratory Tract Infection with and Without a History of Allergy: Age and Gender Aspects

It is well established that the formation of serum proteins, including immunoglobulins, alters with increasing age, a feature that is particularly outstanding in children leading to setting different age-depending reference levels (Garcia-Prat et al.

Table 2 Immunoglobulin G changes distributed by age groups in children with recurrent respiratory tract infections without a history of allergy (RTI-alone)

IgG subclass	≤ 6 years old n (%)			6.5–12 years old n (%)			12.5–18 years old n (%)		
	Decreased	**Normal**	Increased	Decreased	**Normal**	Increased	Decreased	**Normal**	Increased
IgG total	37 (14)	**204 (77)**	25 (9)	8 (8)	**75 (79)**	12 (13)	3 (9)	**27 (82)**	3 (9)
IgG1	66 (25)	**194 (73)**	6 (2)	26 (27)	**62 (65)**	7 (8)**	16 (48)*	**17 (52)**	0 (0)
IgG2	13 (5)	**244 (92)**	9 (3)	7 (7)	**84 (89)**	4 (4)	4 (12)	**29 (88)**	0 (0)
IgG3	39 (15)	**225 (84)**	2 (1)	17 (18)	**75 (79)**	3 (3)	14 (42)*	**19 (58)**	0 (0)
IgG4	54 (20)	**200 (75)**	12 (5)	16 (17)	**64 (67)**	15 (16)	9 (27)	**16 (49)**	8 (24)**

Normal range in each age group **in bold**; *$p < 0.05$ against the corresponding decreases in the other age groups; **$p < 0.05$ against the corresponding increases in the other age groups (Chi^2 tests)

Table 3 Immunoglobulin G changes distributed by age groups in children with recurrent respiratory tract infections with a history of allergy (RTI-allergy)

IgG subclass	≥ 6 years old n (%)			6.5–12 years old n (%)			12.5–18 years old n (%)		
	Decreased	**Normal**	Increased	Decreased	**Normal**	Increased	Decreased	**Normal**	Increased
IgG total	12 (18)	**51 (75)**	5 (7)	3 (7)	**29 (69)**	10 (24)	2 (10)	**15 (75)**	3 (15)
IgG1	20 (29)	**48 (71)**	0 (0)	11 (26)	**27 (64)**	4 (10)**	9 (45)*	**10 (50)**	1 (5)
IgG2	3 (4)	**63 (93)**	2 (3)	3 (7)	**35 (83)**	4 (10)	2 (10)	**17 (85)**	1 (5)
IgG3	7 (10)	**60 (88)**	1 (2)	10 (24)	**32 (76)**	0 (0)	8 (40)*	**11 (55)**	1 (5)
IgG4	11 (16)	**52 (77)**	5 (7)	3 (7)	**27 (64)**	12 (29)	5 (25)*	**7 (35)**	8 (40)**

Normal range in each age group **in bold**; *$p < 0.05$ against the corresponding decreases in the other age groups; **$p < 0.05$ against the corresponding increases in the other age groups (Chi^2 tests)

2018). Therefore, in further evaluation of immunoglobulin G alterations, we divided the cohort of children investigated into three age groups: ≤6 years old, 6.5–12 years old, and 12.5–18 years old. The results of this evaluation in both RTI-alone and RTI-allergy groups are displayed in Tables 2 and 3, respectively. Immunoglobulin G alterations basically went along the same line in both groups. A significantly greater percentage of children showed decreases in IgG1, IgG3, and IgG4 in the oldest children compared with each of the two younger groups. Interestingly, we also identified a significant proportion of children in the oldest group only that displayed significant increases in IgG4 above the reference level.

Concerning gender, we found that male children had a significantly greater propensity for deficiencies in total IgG and in IgG1 and IgG3 subclasses ($p < 0.05$). This effect concerned male children, across the whole spectrum of age studied, but in the group of RTI-alone. The effect was unnoticed in male children with respiratory tract infections and a history of allergy.

4 Discussion

The rational for this study was that allergy in general, and allergic asthma in particular, are known to increase the likelihood of immunoglobulin disorders, notably of the IgG class, leading to increased propensity for recurrent respiratory tract infections (Scott-Taylor et al. 2018; Wahn and von Bernuth 2017). Therefore, it seemed warranted to compare profiles of IgG immunoglobulin alterations between children suffering from frequent respiratory infections with and without a history of allergy. We found that the 65–90% majority of children in both RTI-alone and RTI-allergy groups had the IgG subclass levels within the normal limit. However, we confirmed that some children hospitalized due to RTI had immunoglobulin G abnormalities, consisting of deficiencies in specific IgG subclasses, namely in IgG1, IgG3, IgG4. In rare cases, deficiency was paired. Overall, deficiency was present in less than one-third of children. However, proportions of deficiencies were similar in both RTI-alone and RTI-allergy groups. A lack of effect of allergy juxtaposed to RTI on immunoglobulin G profile is in line with studies that have reported no association between serum IgG glycosylation, an essential regulatory element of immunofunction, and allergic diseases in children (Pezer et al. 2016).

The only significant effect in the RTI-allergy group we noticed was a modestly smaller proportion of children with IgG4 deficiency than that in the RTI-alone group (Fig. 1). This kind of sparing IgG4 from becoming deficient could plausibly reflect the known inhibitory role of IgG4 in IgE-mediated allergic reactions (van de Veen et al. 2013). Allergic asthma, sorted out as a specific subgroup of children with RTI, did not enhance the proportion of IgG deficiency. The grossly unaltered profile of immunoglobulin G deficiency in children with allergic versus non-allergic RTI does not exclude the presence of background allergy-related differences preceding RTI onset. That, however, could not be tested in this study due to its retrospective nature.

We further demonstrate that immunoglobulin G deficiencies were age-dependent in that their number significantly increased with children's age. Male gender also appeared to increase the propensity to immunoglobulin G deficiency; the finding was limited only to the RTI-alone group. Interestingly, we also found a reverse abnormality, i.e., an increase in IgG4 with increasing age. These findings require further exploration using alternate study designs.

Immunoglobulins of the IgG class constitute the most important and largest pool of antibodies in humans (about 60–75% of all immunoglobulins). They are one of the most effective ways to eliminate pathogens in the so-called secondary response, which has a protective effect against infections with encapsulated bacteria (*Haemophilus influenzae, Streptococcus pneumoniae, Neisseria meningitidis*), *Moraxella, Staphylococcus aureus,* and against enteroviral infections. The immunoglobulins have the ability to activate complement proteins and bind to Fc receptor protein on the surface of cells in the immune system, making the process of pathogen elimination far more efficient. Depending on the type of heavy chain, four IgG subclasses have been isolated, whose content alters with age (Kutukculer et al. 2007). Hence, standards levels for children and adults have been developed (Gregorek et al. 1994). Formation of each individual IgG subclass is regulated independently of the others. The content of IgG subclasses also is influenced by ongoing infection which, depending on the type of antigenic material (polysaccharide, protein, or carbohydrate), stimulates the formation of antibodies in various subclasses to a different extent. It has been shown that exposure to protein antigens leads to an increase in IgG1 and less so in IgG3 and IgG4. Exposure to polysaccharide antigens causes a notably increase in IgG2. When exposure is prolonged, the content of all four IgG subclasses may increase, with their normalization taking place after recovery. In the course of some infections, IgG antibodies do not play a significant role in pathogen elimination, for instance, *Mycoplasma, Chlamydia*, or leprosy genus, but they may still be a good diagnostic tool in pathogen identification and monitoring the course of infection (Rastawicki et al. 2009).

The understanding of immunoglobulin function and importance is still limited as about 20% of healthy people show a variable deficiency of IgG subclasses, apparently without any clinical significance (Fried and Bonilla 2009; Tabatabaie et al. 2008). In other cases, IgG subclass deficiency may result in recurrent or chronic respiratory infections, often accompanied by a tendency to develop bronchiectasis (Ozkan et al. 2005). For the first time, deficiency of IgG2 and IgG4 was reported by Oxelius in 1974, with concomitant deficiency of polysaccharide antibodies against *Haemophilus influenzae* in two siblings with persistent ear infections and pneumonia. A different study reported 20 children suffering from asthma, with recurrent respiratory infections, such as sinusitis, bronchitis, and pneumonia, 60% of whom had a selective IgG2 deficiency and 6% had a concomitant IgG2 and IgG3 deficiency (Umetsu et al. 1985). Those with selective IgG2 deficiency poorly responded to the polysaccharide vaccine against *Haemophilus influenzae*, despite normal responses to protein antigens. Likewise, a notable deficiency of IgG2, with otherwise normal total IgG content, has been reported in a group of Dutch children who were hospitalized twice due to bronchitis or pneumonia. About 10% of them were diagnosed with bronchiectasis at the time of admission to hospital (Schatorjé et al. 2016). Selective IgG1 deficiency has also been repeatedly reported in children and adults alike suffering from respiratory tract infections. In case of adults, the deficiency appears often linked to insufficient response to polyvalent pneumococcal polysaccharide vaccination, asthma and atopy, multiple drug allergies, and a host of autoimmune disorders, such as rheumatoid arthritis, psoriasis, Grave's disease, temporal arteritis, and polyneuritis, and others (Barton et al. 2016; Abrahamian et al. 2010; Vanlerberghe et al. 2006; Chee et al. 2001; van Kessel et al. 1999). Deficiency of antibodies contained in IgG subclasses may predispose to the presence of nasal mucosa polyps (Tran Khai Hoan et al. 2014) or chronic sinusitis (Dine et al. 2017; Abrahamian et al. 2009). Deficits of IgG subclasses in different constellations have also been reported in children with chronic obstructive pulmonary disease (Leitao Filho et al. 2018; Bergera et al. 2017).

In conclusion, we found that a great majority of children with recurrent respiratory tract infections had the immunoglobulin G levels within the normal limit. The remaining fraction of children had variable deficiencies of specific IgGs, most often affecting IgG1, IgG3, and IgG4; frequency of deficiencies ranging from about 10% to 40%. Immunoglobulin deficiency was increasing with children's age and tended to be more frequent in male gender. Importantly, we failed to substantiate the existence of any distinct abnormality(s) in the immunoglobulin G profile which would be characteristic to a clinical history of allergy accompanying recurrent RTI in children. Therefore, the measurement of IgGs could hardly be of help in the differential diagnostics of the allergic background of RTI. The exact role of immunoglobulin G in preventing recurrent respiratory tract infections is still an area of limited understanding as long as it remains unclear why some children with IgG subclass deficiency suffer from persistent respiratory infections and others do not. The corollary is that the therapeutic supplementation of immunoglobulins as a preventive or curative measure for recurrent respiratory infections, in particular in case of a history of allergy, although it is used in some instances with success (Kim et al. 2017; Parker et al. 2017; Olinder-Nielsen et al. 2007), requires further exploratory research.

Conflicts of Interest The authors declare no conflicts of interest in relation to this article.

Ethical Approval All procedures performed in studies involving human participants were in accordance with the ethical standards of the institutional and/or national research committee and with the 1964 Helsinki declaration and its later amendments or comparable ethical standards.

Informed Consent A retrospective nature of the study consisting of an anonymous review of medical files obviated the need to obtain consent from all individual participants included in the study.

References

Abrahamian F, Agrawal S, Gupta S (2009) Immunological and clinical profile of adult children with selective immunoglobulin subclass deficiency: response to intravenous immunoglobulin therapy. Clin Exp Immunol 159:344–350

Abrahamian F, Agrawal S, Gupta S (2010) IgG3 deficiency: common in obstructive lung disease. Hereditary in families with immunodeficiency and autoimmune disease. Clin Exp Immunol 159(3):344–350

Barton JC, Bertoli LF, Barton JC, Acton RT (2016) Selective subnormal IgG1 in 54 adult index patients with frequent or severe bacterial respiratory tract infections. J Immunol Res 2016:1405950

Bergera M, Gengb B, Cameronc DW, Murphya LM, Schulmand ES (2017) Primary immune deficiency diseases as unrecognized causes of chronic respiratory disease. Respir Med 132:181–188

Chee L, Graham SM, Carothers DG, Ballas ZK (2001) Immune dysfunction in refractory sinusitis in a tertiary care setting. Laryngoscope 111(2):233–235

Dine G, Ali-Ammar N, Brahimi S, Rehn Y (2017) Chronic sinusitis in a patient with selective IgG4 subclass deficiency controlled with enriched immunoglobulins. Clin Case Rep 5(6):792–794

Falconer AE, Carson R, Johnstone R, Bird P, Kehoe M, Calvert JE (1993) Distinct IgG1 and IgG3 subclass responses to two streptococcal protein antigens in man: analysis of antibodies to streptolysin O and M protein using standardized subclass-specific enzyme-linked immunosorbent assays. Immunology 79(1):89–94

Fried AJ, Bonilla FA (2009) Pathogenesis, diagnosis, and management of primary antibody deficiencies and infections. Clin Microbiol Rev 22(3):396–414

Garcia-Prat M, Vila-Pijoan G, Martos Gutierrez S, Gala Yerga G, García Guantes E, Martínez-Gallo M, Martín-Nalda A, Soler-Palacín P, Hernández-González M (2018) Age-specific pediatric reference ranges for immunoglobulins and complement proteins on the Optilite™ automated turbidimetric analyzer. J Clin Lab Anal 32(6):e22420

Gregorek H, Imielska D, Gornicki J, Mikołajewicz J, Przeradzka B, Madaliński K (1994) Development of IgG subclasses in healthy polish children. Arch Immunol Ther Exp 42(5–6):377–382

Kim JH, Ye YM, Ban GY, Shin YS, Lee HY, Nam YH, Lee SK, Cho YS, Jang SH, Jung KS, Park HS (2017) Effects of immunoglobulin replacement on asthma exacerbation in adult asthmatics with IgG subclass deficiency. Allergy Asthma Immunol Res 9(6):526–533

Kutukculer N, Karaca NE, Demircioglu O, Aksu G (2007) Increases in serum immunoglobulins to age-related normal levels in children with IgA and/or IgG subclass deficiency. Pediatr Allergy Immunol 18(2):167–173

Leitao Filho FS, Ra SW, Mattman A, Schellenberg RS, Criner GJ, Woodruff PG, Lazarus SC, Albert R, Connett JE, Han MK, Martinez FJ, Leung JM, Paul Man SF, Aaron SD, Reed RM, Sin DD, Canadian Respiratory Research Network (CRRN) (2018) Serum IgG subclass levels and risk of exacerbations and hospitalizations in children with COPD. Respir Res 19(1):30

Michaelsen TE, Sandlie I, Bratlie DB, Sandin RH, Ihle O (2009) Structural difference in the complement activation site of human IgG1 and IgG3. Scand J Immunol 70 (6):553–564

Olinder-Nielsen AM, Granert C, Forsberg P, Friman V, Vietorisz A, Bjorkander J (2007) Immunoglobulin prophylaxis in 350 adults with IgG subclass deficiency and recurrent respiratory tract infections: a long-term follow-up. Scand J Infect Dis 39:44–50

Oxelius VA (1974) Chronic infections in a family with hereditary deficiency of IgG2 and IgG4. Clin Exp Immunol 17:19–27

Ozkan H, Atlihan F, Genel F, Targan S, Gunvar T (2005) IgA and/or IgG subclass deficiency in children with recurrent respiratory infections and its relationship with chronic pulmonary damage. J Investig Allergol Clin Immunol 15(1):69–74

Parker AR, Skold M, Ramsden DR, Ocejo-Vinyals JG, López-Hoyos M, Harding S (2017) The clinical utility of measuring IgG subclass immunoglobulins during immunological investigation for suspected primary antibody deficiencies. Lab Med 48(4):314–325

Pezer M, Stambuk J, Perica M, Razdorov G, Banic I, Vuckovic F, Gospic AM, Ugrina I, Vecenaj A, Bakovic MP, Lokas SB, Zivkovic J, Plavec D, Devereux G, Turkalj M, Lauc G (2016) Effects of allergic diseases and age on the composition of serum IgG glycome in children. Sci Rep 6:33198

Rastawicki W, Rokosz N, Jagielski M (2009) Subclass distribution of human igg antibodies to mycoplasma pneumoniae in the course of mycoplasmosis. Med Dosw Mikrobiol 61(4):375–379. (Article in Polish)

Schatorjé EJ, de Jong E, van Hout RW, García Vivas Y, de Vries E (2016) The challenge of immunoglobulin-G subclass deficiency and specific polysaccharide antibody deficiency – a Dutch Pediatric Cohort Study. J Clin Immunol 36(2):141–148

Scott-Taylor TH, Axinia SC, Amin S, Pettengell R (2018) Immunoglobulin G; structure and functional implications of different subclass modifications in initiation and resolution of allergy. Immun Inflamm Dis 6 (1):13–33

Tabatabaie P, Aghamohammadi A, Mamishi S, Isaeian A, Heidari G, Abdollahzade S, Pirouzi P, Rezaei N, Heidarnazhad H, Mirsaeid Ghazi B, Yeganeh M, Cheraghi T, Abolhasani H, Saghafi S, Alizadeh H, Anaraki MR (2008) Evaluation of humoral immune function in children with bronchiectasis. Iran J Allergy Asthma Immunol 7(2):69–77

Tran Khai Hoan N, Karmochkine M, Laccourreye O, Bonfilsa P (2014) Nasal polyposis and immunoglobulin-G subclass deficiency. Eur Ann Otorhinolaryngol Head Neck Dis 131(3):171–175

Umetsu DT, Ambrosino DM, Quinti I, Siber GR, Geha RS (1985) Recurrent sinopulmonary infection and impaired antibody response to bacterial capsular polysaccharide antigen in children with selective IgG-subclass deficiency. N Engl J Med 313 (20):1247–1251

van de Veen W, Stanic B, Yaman G, Wawrzyniak M, Söllner S, Akdis DG, Rückert B, Akdis CA, Akdis M (2013) IgG4 production is confined to human IL-10-producing regulatory B cells that suppress antigen-specific immune responses. J Allergy Clin Immunol 131:1204–1212

Van Kessel DA, Horikx PE, Van Houte AJ, De Graaff CS, Van Velzen-Blads H, Rijkers GT (1999) Clinical and immunological evaluation of patients with mild IgG1 deficiency. Clin Exp Immunol 118(1):102–107

Vanlerberghe L, Joniau S, Jorissen M (2006) The prevalence of humoral immunodeficiency in refractory rhinosinusitis: a retrospective analysis. B-ENT 2:161–166

Vidarsson G, Dekkers G, Rispens T (2014) IgG subclasses and allotypes: from structure to effector functions. Front Immunol 5:520

Wahn V, von Bernuth H (2017) IgG subclass deficiencies in children: facts and fiction. Pediatr Allergy Immunol 28(6):521–524

Adv Exp Med Biol - Clinical and Experimental Biomedicine (2021) 11: 71–77
https://doi.org/10.1007/5584_2020_548

Published online: 27 June 2020

Pruritus Characteristics in Severe Atopic Dermatitis in Adult Patients

Andrzej Kazimierz Jaworek, Krystyna Szafraniec, Zbigniew Doniec, Magdalena Jaworek, Anna Wojas-Pelc, and Mieczysław Pokorski

Abstract

Atopic dermatitis (AD) is classified as a most common inflammatory skin disease. The condition is characterized by recurrent eczematous lesions and intense pruritus or itch, a hallmark of AD. The aim of this study was to identify the provoking factors of itch in severe AD adult patients. There were 34 adult patients suffering from AD of the median age of 40 years enrolled into the study and a control group that consisted of 20 healthy subjects. The severity of AD was assessed with the SCORing Atopic Dermatitis (SCORAD) index, pruritus intensity was assessed on a visual analog scale (VAS), and itch aggravating factors were assessed with a questionnaire. Specific IgE (sIgE) antibodies and interleukin IL-33 were measured in venous blood. We found that all the patients with severe AD had intensive itch (VAS: 9–10 points) during the whole day and 30 (88.2%) patients had it during nighttime. The most significant factors aggravating itch were the following: dry skin (27 patients; 79.4%), exposure to dust mite (22 patients; 64.7%,), and emotional distress (17 patients; 50%). Moreover, there was a positive correlation between the intensity of itch and the level of sIgE antibodies to dust mite ($p < 0.001$). The content of IL-33 was significantly higher in AD patients with severe skin lesions. This exploratory study shows that skin dryness, dust exposure, and distress play an essential role in the exacerbation of AD in the adult population.

Keywords

Atopic dermatitis · Emotional distress · IgE antibody · Inflammation · Itching · Pruritus · Skin

A. K. Jaworek (✉) and A. Wojas-Pelc
Department of Dermatology, Jagiellonian University Medical College, Cracow, Poland
e-mail: andrzej.jaworek@uj.edu.pl

K. Szafraniec
Department of Epidemiology and Population Studies, Institute of Public Health, Faculty of Health Sciences, Jagiellonian University Medical College, Cracow, Poland

Z. Doniec
Department of Pneumology, Institute of Tuberculosis and Lung Disorders, Field Unit in Rabka, Rabka, Poland

M. Jaworek
Department of Physiotherapy, Faculty of Health Sciences, Jagiellonian University Medical College, Cracow, Poland

M. Pokorski
Institute of Health Sciences, Opole Medical School, Opole, Poland

1 Introduction

Atopic dermatitis (AD) is classified as a most common inflammatory skin disease in the

Table 1 Characteristics of atopic dermatitis (AD) patients and control subjects

		Patients ($n = 34$)	Controls ($n = 20$)
Gender: n (%)	Male	16 (47.1)	10 (50)
	Female	18 (52.9)	10 (50)
Age: median; min-max (years)		40 (20–70)	27 (23–72)
SCORAD: median; min-max (points)		60.6 (50.2–80.4)	0 (0–0)
VAS: n (%) (points)	10	29 (85.3)	0 (100)
	9	5 (14.7)	0 (100)

SCORAD SCORing atopic dermatitis index, *VAS* visual analogue scale

developed world, with a lifetime prevalence of 15–20%. The condition is characterized by recurrent eczematous lesions such as erythema, papules, and vesicles in acute stage; lichenification; and scaling and fissuring at later stages. Above all, intense pruritus or itch is the hallmark of AD. The phenotypes of the disease are extremely variable, ranging from localized small lesions at the age-dependent predilection sites to erythroderma, an entire body manifestation (Weidinger et al. 2018). Pruritus is the most essential diagnostic feature of AD, affecting social and family life and reducing quality of life (Silverberg et al. 2020). Despite several studies, the pathophysiology of itch and its aggravating factors in AD are yet not full well known (Fujii 2020). A recent study has identified the epithelial cell-derived cytokine IL-33 (interleukin 33) as an upstream factor of type 2 cellular responses. IL-33 is responsible for activation of many a cell in the AD skin and stimulates itch-sensory neurons (Klonowska et al. 2018). Therefore, the aim of this study was to identify the factors provoking itch in adult patients with severe AD. We addressed the issue using a visual analogue scale (VAS) for pruritus severity in connection with the assessment of serum content of specific immunoglobulin E (sIgE) and IL-33 cytokine.

2 Methods

2.1 Patients and Skin Lesions

This study was performed in 34 adult patients suffering from AD (F/M: 18/16; median age: 40 years old; range: 20–70 years). The AD was diagnosed according to the Hanifin and Rajka (1980) criteria. The severity of AD was based on the SCORing Atopic Dermatitis (SCORAD) index, where score over 50 points indicates a severe form of the disease (Oranje et al. 2007). All the patients were classified as having severe AD (median SCORAD index: 60.6 points; min-max: 50.2–80.4 points). Further, each patient performed a self-assessment of the severity of the most intensive pruritus during the preceding 24 h using a Visual Analogue Scale (VAS), where 0 points was none, 1–3 was mild pruritus, 4–6 was moderate pruritus, 7–8 was severe pruritus, and 9–10 was very severe pruritus (Reich et al. 2017). Basic characteristics of patients were presented in Table 1. All the patients were asked to assess five most important itch-aggravating factors, based on a short questionary listing fifteen collated from recent publications and from clinical practice (Jeziorkowska et al. 2019; Chrostowska-Ptak et al. 2009; Dawn et al. 2009; Langan and Williams 2006; Wahlgren 1991). The control group consisted of 20 healthy volunteers, gender- and age-matched (F/M: 10/10; median age: 27 years; min-max: 23–72 years).

2.2 Biochemical Indices

Blood samples were taken from ulnar venipuncture in a volume of 10 mL between 7:00 am and 8:00 am. The samples were kept for 2 h at room temperature for a clot to form, then centrifuged at 3500 rpm for 10 min, and frozen at −80 °C until use. Dust mite-sIgE (d2), grass sIgE (g6), trees sIgE (t3), and weed sIgE (w6) antibodies were assessed using a UniCAP 100 fluoro-enzymatic immunoassay (FEIA) (ImmunoCAP-System;

Phadia AB, Sweden) according to manufacturer's instructions. The level of IgE antibodies was classified as follows: Class I: 0.35–0.69 IU/mL, Class II: 0.70–3.49 IU/mL, Class III: 3.50–17.49 IU/mL, Class IV: 17.50–49.90 IU/mL, Class V: 50.00–100.00 IU/mL, and Class VI: >100 IU/mL. The cytokine IL-33 was assessed using an ELISA assay (R&D System, Minneapolis, MN).

Nominal and ordinal data were given as counts and percentages, and they were expressed as medians and minimum-maximum values. The Mann-Whitney rank sum test and Fisher exact test were used for intergroup statistical comparisons of non-parametric variables. A p-value <0.05 defined statistically significant differences.

3 Results

Nearly all the patients, 32 (94.1%), had been suffering from AD for over 10 years, and 30 (88.2%) patients declared a constantly persisting exacerbation. All the patients evaluated pruritus as being very intense (VAS: 9 and 10 points). Thirty patients reported that pruritus was presented during the whole day. Only did four patients report intense pruritus just during the evening hours. The characteristics of pruritus are presented in Table 2. Most often pruritus was aggravated by dry skin (27 patients; 79.4%), contact with dust (22 patients; 64.7%), and emotional distress (17 patients; 50%) (Fig. 1). The level of sIgE antibodies to dust mite most frequently

Table 2 Characteristics of pruritus in atopic dermatitis (AD) patients

		Patients ($n = 34$)
Number of exacerbations: n (%)	>5 *per* month	4 (11.8)
	Constantly persisting	30 (88.2)
Duration of pruritus: n (%)	<1 day	4 (11.8)
	Constantly persisting	30 (88.2)
Intermittent night pruritus: n (%)	Yes	33 (97.1)
	No	1 (2.9)
Maximum pruritus day and: n (%)	Whole day	30 (88.2)
	Evening	4 (11.8)

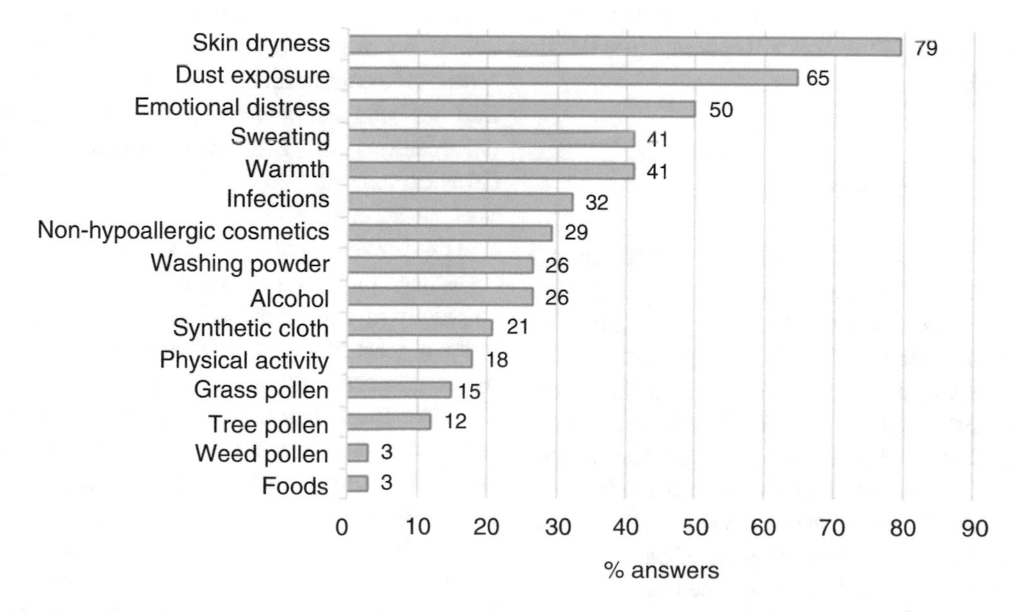

Fig. 1 Pruritus aggravating factors in adult patients suffering from atopic dermatitis (AD) ($n = 34$)

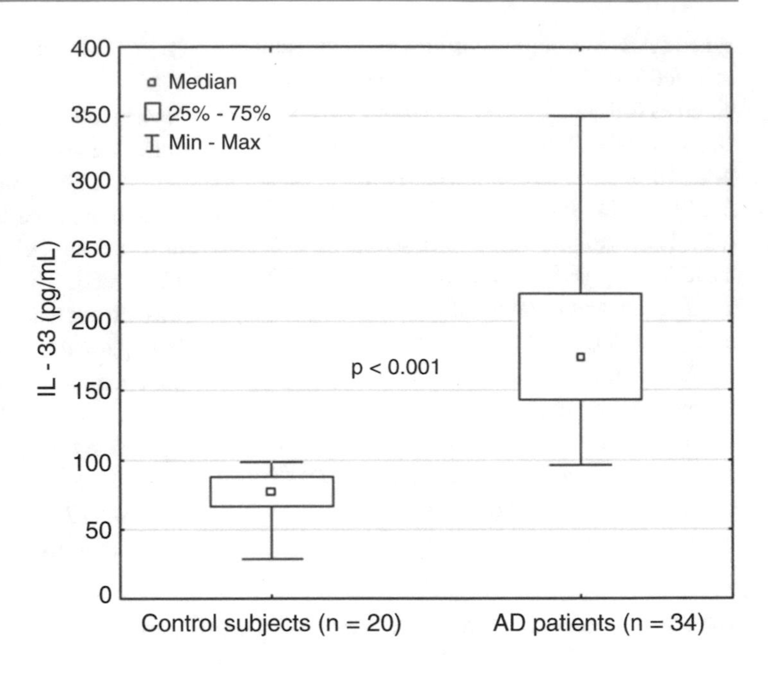

Fig. 2 Serum IL-33 in adult patients suffering from atopic dermatitis (AD) and in control subjects

corresponded to Class V or VI, second to last and last highest, in patients with dust contact as an aggravating factor of pruritus exacerbations when compared to other patients (95.5% vs. 33.3%, $p < 0.001$). There were no significant relationships between other sIgE antibodies and the corresponding factors aggravating pruritus. Blood content of IL-33 was significantly higher in the AD patients (median 173.4 pg/mL) than in the control subjects (median 77.0 pg/mL) ($p < 0.001$) (Fig. 2).

4 Discussion

Recent studies indicate a rather widespread AD morbidity not only in the pediatric population but among adults as well (Barbarot et al. 2018). In 2019, a consensus for AD management in adult patients was published under the affiliation of the Australian Dermatological Association (Smith et al. 2020). Chronic pruritus is a basic sign of AD, which is commonly described as "the itch that rashes" (Murota and Katayama 2017; Romeo 1995). In the extremely complicated pathophysiology of atopic itch, the crucial issue is a differentiation between the canonical pruritogen histamine-dependent and histamine-independent itch. The latter is related to the proteins belonging to the cytokine family, such as interleukin (IL)-2, IL-4, IL-13, IL-31, IL-33, and thymic stromal lymphoprotein (TSLP) (Cevikbas and Lerner 2020; Mollanazar et al. 2016). The importance of pruritus assessment in gauging the severity of AD has been recently recognized in the literature (Huet et al. 2019; O'Neill et al. 2011; Chrostowska-Ptak et al. 2009). According to the European, Polish, and Japanese Recommendations for AD treatment, pruritus has been distinguished as the main trigger of exacerbations of skin manifestation of AD (Nowicki et al. 2020; Katoh et al. 2019; Wollenberg et al. 2018).

The present study was unique in that we investigated the patients suffering from AD with extremely intense pruritus, having the SCORAD index at least 50 or more. Nearly all the patients reported intermittent sleep disruption due to pruritus. We found that pruritus commonly arose in response to skin dryness, contact with dust mite, and in emotional distress. Skin dryness as the most significant prognostic of AD progression has been noted by Chrostowska-Ptak et al. (2009). On the other hand, exposure to dust mite has been reported by Langan and Williams (2006) as a trigger of exacerbations in children with AD. Skin dryness and emotional distress have been found to trigger exacerbations in 93% and 63% of AD patients, respectively, in the French population (Brenaut et al. 2013). AD is commonly accompanied by psychiatric problems (Thyssen et al. 2018). A connection between emotional distress and pruritus has self-perpetuating implications, although which is the primary causative factor has yet to be identified (Sanders and Akiyama 2018). It also is considered that aeroallergens belong to essential triggers of exacerbations in adult patients suffering from AD (Nowicki et al. 2020). Like in the present study, Jeziorkowska et al. (2019) have reported that one of the most important triggers of pruritus exacerbations is the dust mite. Contact with dust mites is often accompanied by enhanced blood level of sIgE antibodies (Scalabrin et al. 1999). Cysteine and serine proteases, the main allergen proteins in dust mite, damage the epidermal barrier, causing a detriment to both innate and acquired immune responses engaged in the pathogenesis of AD (Miller 2019). Cat hair has been indicated as the essential pruritic factor in a Chinese study encompassing 396 patients with AD (Hu et al. 2020). Likewise, aeroallergens coming from guinea pig or dog coat have been highlighted in other studies (Jaworek et al. 2019, 2020).

In this study, we found a significant increase in blood content of IL-33 in AD patients when compared with healthy control subjects. The finding is in line with similar reports in the literature (Wang et al. 2020; Nygaard et al. 2016; Tamagawa-Mineoka et al. 2014). IL-33, immunomodulatory cytokine, is a member of the IL-1 family of cytokines, which are also known as the epithelial cell-derived cytokines or alarmins. A spate of cells such as keratinocytes, adipocytes, or endothelial cells produce IL-33 in response to mechanical injury of the epithelial barrier due to scratching. IL-33 binds to interleukin 1 receptor-like 1, also known as IL1RL1 and ST2 receptors, which are expressed on inflammatory cells, such as basophils and mastocytes, promoting Th2-type immune responses (Klonowska et al. 2018). IL-33 is a proinflammatory cytokine activating innate lymphomatoid cells 2 (ILC2) and stimulating Th2-dependent IL-5 and IL-13 cytokine responses, without antigen contact. Imai et al. (2019) have demonstrated that IL-33, upregulated in keratinocytes of transgenic mice line, induces severe eczema and itching in many ways, inter alia by histamine release from mastocytes and basophils. Histamine acts on sensory skin receptors promoting the itch. Scratching behavior arises that perpetuate the release of histamine and cytokines from keratinocytes, which exacerbates the itching, leading to a self-driven mechanism of itching (Imai 2019; Dong and Dong 2018). Etokimab, a humanized IgG1 antibody that targets IL-33, offers promise as effective treatment of pruritus in AD and currently is in phase 2a trial. A rapid improvement in the sensation of itch has been noticed after a single systemic administration of etokimab in a group of 12 patients with AD (Chen et al. 2019).

In conclusion, we believe we have shown that skin dryness, exposure to dust mite, and emotional distress are the most common factors inducing intense itch in patients with severe atopic dermatitis, who suffered from unrelenting skin symptoms during the day and night, including itch-related intermittent sleep disruptions. Particularly, blood sIgE antibodies to dust mite outstandingly correlated with the intensity of itch. We further demonstrate an increase in blood IL-33, a pro-inflammatory, itch-promoting cytokine. This report summarizes our practical knowledge concerning the causative factors of exacerbations of skin lesions in adult atopic dermatitis, pointing to possible preventive interventions. The exact mechanisms of the disease remain enigmatic. However, progress in unraveling novel biomarkers involved in the molecular pathomechanisms of atopic dermatitis, particularly linked to interleukin-1 family of cytokines, opens new avenues of research with possible benefits of more effective targeted therapeutic approaches.

Conflicts of Interest The authors declare no conflicts of interest in relation to this chapter.

Ethical Approval All procedures performed in studies involving human participants were in accordance with the ethical standards of the institutional and/or national research committee and with the 1964 Helsinki declaration and its later amendments or comparable ethical standards. The study was approved by the Jagiellonian University Ethics Committee.

Informed Consent Written informed consent was obtained from all individual participants included in the study.

References

Barbarot S, Auziere S, Gadkari A, Girolomoni G, Puig L, Simpson EL, Margolis DJ, de Bruin-Weller M, Eckert L (2018) Epidemiology of atopic dermatitis in adults: results from an international survey. Allergy 73 (6):1284–1293

Brenaut E, Garlantezec R, Talour K, Misery L (2013) Itch characteristics in five dermatoses: non-atopic eczema, atopic dermatitis, urticaria, psoriasis and scabies. Acta Derm Venereol 93(5):573–574

Cevikbas F, Lerner EA (2020) Physiology and pathophysiology of itch. Physiol Rev 100(3):945–982

Chen YL, Gutowska-Owsiak D, Hardman CS, Westmoreland M, MacKenzie T, Cifuentes L, Waithe D, Lloyd-Lavery A, Marquette A, Londei M, Ogg G (2019) Proof-of-concept clinical trial of etokimab shows a key role for IL-33 in atopic dermatitis pathogenesis. Sci Transl Med 11(515):pii: eaax2945

Chrostowska-Ptak D, Salomon J, Reich A, Szepietowski JC (2009) Clinical aspects of itch in adult atopic dermatitis patients. Acta Derm Venereol 89(4):379–383

Dawn A, Papoiu AD, Chan YH, Rapp SR, Rassette N, Yosipovitch G (2009) Itch characteristics in atopic dermatitis: results of a web-based questionnaire. Br J Dermatol 160(3):642–644

Dong X, Dong X (2018) Peripheral and central mechanisms of itch. Neuron 98(3):482–494

Fujii M (2020) Current understanding of pathophysiological mechanisms of atopic dermatitis: interactions among skin barrier dysfunction, immune abnormalities and pruritus. Biol Pharm Bull 43(1):12–19

Hanifin JM, Rajka G (1980) Diagnostic features of atopic dermatitis. Acta Dermatol Venerol (Stockh) 92 (Suppl):44–47

Hu Y, Liu S, Liu P, Mu Z, Zhang J (2020) Clinical relevance of eosinophils, basophils, serum total IgE level, allergen-specific IgE, and clinical features in atopic dermatitis. J Clin Lab Anal. https://doi.org/10.1002/jcla.23214

Huet F, Faffa MS, Poizeau F, Merhand S, Misery L, Brenaut E (2019) Characteristics of pruritus in relation to self-assessed severity of atopic dermatitis. Acta Derm Venereol 99(3):279–283

Imai Y (2019) Interleukin-33 in atopic dermatitis. J Dermatol Sci 96(1):2–7

Imai Y, Yasuda K, Nagai M, Kusakabe M, Kubo M, Nakanishi K, Yamanishi K (2019) IL-33-induced atopic dermatitis-like inflammation in mice is mediated by group 2 innate lymphoid cells in concert with basophils. J Invest Dermatol 139(10):2185–2194

Jaworek AK, Szafraniec K, Jaworek M, Doniec Z, Zalewski A, Kurzawa R, Wojas-Pelc A, Pokorski M (2019) Cat allergy as a source intensification of atopic dermatitis in adult patients. Adv Exp Med Biol 1251:39–47

Jaworek AK, Szafraniec K, Jaworek M, Hałubiec P, Wojas-Pelc A (2020) Is the presence of a fur animal an exacerbating factor of atopic dermatitis in adults? Pol Merkur Lekarski 48(283):19–22

Jeziorkowska R, Rożalski M, Skowroński K, Samochocki Z (2019) Can evaluation of specific immunoglobulin E serum concentrations of antibodies to aeroallergens in atopic dermatitis patients replace skin prick tests method in clinical practice? Adv Dermatol Allergol 36(4):478–484

Katoh N, Ohya Y, Ikeda M, Ebihara T, Katayama I, Saeki H, Shimojo N, Tanaka A, Nakahara T, Nagao M, Hide M, Fujita Y, Fujisawa T, Futamura M, Masuda K, Murota H, Yamamoto-Hanada K (2019) Clinical practice guidelines for the management of atopic dermatitis 2018. J Dermatol 46 (12):1053–1101

Klonowska J, Gleń J, Nowicki RJ, Trzeciak M (2018) New cytokines in the pathogenesis of atopic dermatitis-new therapeutic targets. Int J Mol Sci 19 (10):3086. https://doi.org/10.3390/ijms19103086

Langan SM, Williams HC (2006) What causes worsening of eczema? A systematic review. Br J Dermatol 155 (3):504–514

Miller JD (2019) The role of dust mites in allergy. Clin Rev Allergy Immunol 57(3):312–329

Mollanazar NK, Smith PK, Yosipovitch G (2016) Mediators of chronic pruritus in atopic dermatitis: getting the itch out? Clin Rev Allergy Immunol 51 (3):263–292

Murota H, Katayama I (2017) Exacerbating factors of itch in atopic dermatitis. Allergol Int 66(1):8–13

Nowicki RJ, Trzeciak M, Kaczmarski M, Wilkowska A, Czarnecka-Operacz M, Kowalewski C, Rudnicka L, Kulus M, Mastalerz-Migas A, Peregud-Pogorzelski J, Sokołowska-Wojdyło M, Śpiewak R, Adamski Z, Czuwara J, Kapińska-Mrowiecka M, Kaszuba A, Krasowska D, Kręcisz B, Narbutt J, Majewski S, Reich A, Samochocki Z, Szepietowski J, Woźniak A (2020) Atopic dermatitis. Interdisciplinary diagnostic and therapeutic recommendations of the Polish Dermatological Society, Polish Society of Allergology, Polish Pediatric Society and Polish Society of Family Medicine. Part I. Prophylaxis, topical treatment and phototherapy. Adv Dermatol Allergol XXXVII(1):1–10

Nygaard U, Hvid M, Johansen C, Buchner M, Fölster-Holst R, Deleuran M, Vestergaard C (2016) TSLP,

IL-31, IL-33 and sST2 are new biomarkers in endophenotypic profiling of adult and childhood atopic dermatitis. J Eur Acad Dermatol Venereol 30 (11):1930–1938

O'Neill JL, Chan YH, Rapp SR, Yosipovitch G (2011) Differences in itch characteristics between psoriasis and atopic dermatitis patients: results of a web-based questionnaire. Acta Derm Venereol 91(5):537–540

Oranje AP, Glazenburg EJ, Wolkerstorfer A, de Waard-van der Spek FB (2007) Practical issues on interpretation of scoring atopic dermatitis: the SCORAD index, objective SCORAD and the three-item severity score. Br J Dermatol 157(4):645–648

Reich A, Chatzigeorkidis E, Zeidler C, Osada N, Furue M, Takamori K, Ebata T, Augustin M, Szepietowski JC, Ständer S (2017) Tailoring the cut-off values of the visual analogue scale and numeric rating scale in itch assessment. Acta Derm Venereol 97(6):759–760

Romeo SP (1995) Atopic dermatitis: the itch that rashes. Pediatr Nurs 21(2):157–163

Sanders KM, Akiyama T (2018) The vicious cycle of itch and anxiety. Neurosci Biobehav Rev 87:17–26

Scalabrin DM, Bavbek S, Perzanowski MS, Wilson BB, Platts-Mills TA, Wheatley LM (1999) Use of specific IgE in assessing the relevance of fungal and dust mite allergens to atopic dermatitis: a comparison with asthmatic and nonasthmatic control subjects. J Allergy Clin Immunol 104:1273–1279

Silverberg JI, Lai JS, Patel K, Singam V, Vakharia PP, Chopra R, Sacotte R, Kantor R, Hsu DY, Cella D (2020) Measurement properties of the PROMIS Itch Questionnaire – itch severity assessments in adults with atopic dermatitis. Br J Dermatol. https://doi.org/10.1111/bjd.18978

Smith S, Baker C, Gebauer K, Rubel D, Frankum B, Soyer HP, Weightman W, Sladden M, Rawlin M, Headley AP, Somerville C, Beuth J, Logan N, Mewton E, Foley P (2020) Atopic dermatitis in adults: an Australian management consensus. Australas J Dermatol 61:23–32

Tamagawa-Mineoka R, Okuzawa Y, Masuda K, Katoh N (2014) Increased serum levels of interleukin 33 in patients with atopic dermatitis. J Am Acad Dermatol 70(5):882–888

Thyssen JP, Hamann CR, Linneberg A, Dantoft TM, Skov L, Gislason GH, Wu JJ, Egeberg A (2018) Atopic dermatitis is associated with anxiety, depression, and suicidal ideation, but not with psychiatric hospitalization or suicide. Allergy 73(1):214–220

Wahlgren CF (1991) Itch and atopic dermatitis: clinical and experimental studies. Acta Derm Venereol Suppl (Stockh) 165:1–53

Wang S, Zhu R, Gu C, Zou Y, Yin H, Xu J, Li W (2020) Distinct clinical feature and serum cytokine pattern of elderly atopic dermatitis in China. J Eur Acad Dermatol Venereol. https://doi.org/10.1111/jdv.16346

Weidinger S, Beck LA, Bieber T, Kabashima K, Irvine AD (2018) Atopic dermatitis. Nat Rev Dis Primers 4 (1):1

Wollenberg A, Barbarot S, Bieber T, Christen-Zaech S, Deleuran M, Fink-Wagner A, Gieler U, Girolomoni G, Lau S, Muraro A, Czarnecka-Operacz M, Schäfer T, Schmid-Grendelmeier P, Simon D, Szalai Z, Szepietowski JC, Taïeb A, Torrelo A, Werfel T, Ring J (2018) Consensus-based European guidelines for treatment of atopic eczema (atopic dermatitis) in adults and children: part I. J Eur Acad Dermatol Venereol 32 (5):657–682

Adv Exp Med Biol - Clinical and Experimental Biomedicine (2021) 11: 79–88
https://doi.org/10.1007/5584_2020_546

Published online: 3 June 2020

Pathophysiological Responses to a Record-Breaking Multi-hour Underwater Endurance Performance: A Case Study

Vittore Verratti, Gerardo Bosco, Vincenzo Zanon, Tiziana Pietrangelo, Enrico Camporesi, Danilo Bondi, and Mieczyslaw Pokorski

Abstract

The "Endless Diving Project Step 36" took place in the harbor waters of the town of Maratea in Italy in September 2014. The goal of the project was an attempt by an experienced male diver, equipped with a wet 7-mm suit and a normal gas tank, to set the world record-breaking of nonstop underwater performance. We studied inflammatory, hematological, and endocrine responses during the extreme condition of the attempt. Venous blood samples were collected at baseline, the day before the attempt; immediately after the return from underwater; then at Day 1, Day 4, and Day 12; and later at Month 1 and Month 41 of follow-up. We found that there was an increase in the content of blood neutrophils, monocytes, and eosinophils and a decrease in lymphocytes at Day 1 and a late increase in basophils at Day 12 after the dive. Inflammatory markers and hematocrit and hemoglobin increased immediately after the dive, dropped at Day 1, and reverted gradually to the control level from Day 4 to Day 12. Serotonin and dopamine decreased, while adrenaline increased at Day 1, gradually recovering in the days of follow-up. Insulin, luteinizing hormone, growth hormone, and prolactin increased, while testosterone, cortisol, 17-β-estradiol, thyroid-stimulating hormone, and adrenocorticotropic hormone decreased at Day 1, with a partial recovery at Day 4. We conclude that the homeostatic response to the extreme, prolonged underwater performance showed signs of psychological and pro-inflammatory stress. The hormonal response reflected an acute testicular insufficiency. These responses resembled those characteristics for ultra-endurance exercise accompanied by vasculitis and dehydration.

V. Verratti (✉)
Department of Psychological, Health and Territorial Sciences, Università "G. d'Annunzio" di Chieti-Pescara, Chieti, Italy
e-mail: vittore.verratti@unich.it

G. Bosco (✉) and V. Zanon
Department of Biomedical Sciences, Università degli Studi di Padova, Padova, Italy

T. Pietrangelo and D. Bondi
Department of Neuroscience, Imaging and Clinical Sciences, University "G. d'Annunzio" Chieti-Pescara, Chieti, Italy

E. Camporesi
University of South Florida College of Medicine, Tampa, FL, USA

M. Pokorski
Institute of Health Sciences, Opole Medical School, Opole, Poland

Keywords

Diving · Extreme homeostasis · Hormones · Inflammation · Underwater performance

1 Introduction

During diving, the circulatory system is stressed by a high level of oxygen partial pressure (PO_2) (Wilmshurst 1998). High PO_2 accentuates oxidative stress and promotes inflammation via redox and oxygen-sensitive transcription factors (Eftedal et al. 2013; Peng et al. 2008). A decrease in hydrostatic pressure during ascent can stimulate bubble formation in blood and tissues, triggering a decompression illness (DCI) (Catchpole and Gersh 1947), which is associated with physiological and pathophysiological responses, notably with increases in intravascular inert gas content and circulating microparticles. These events promote an increase in the circulating inflammatory molecules and neutrophil and endothelial activation (Thom et al. 2011, 2013; Yang et al. 2012). Eftedal et al. (2013) have demonstrated that a 3-day series of daily dives to 18 m depth for 47 min each, while breathing compressed air, may cause a pronounced change in transcription patterns of specific leukocytes, with downregulation of genes expressed by CD8 + T lymphocytes and neutrophil killer cells and upregulation of genes expressed by neutrophils, monocytes, and macrophages. These results indicate that sublethal oxidative stress elicits the myeloid immune response in scuba diving and that extensive diving may cause persistent changes in pathways controlling apoptosis and inflammation.

In general, systemic inflammation changes the number of circulating white blood cells and platelets and increases the content of C-reactive protein (CRP) (Grivennikov et al. 2010; Sparmann and Bar-Sagi 2004; Yoshida et al. 2002). It would be reasonable to suppose that a 36-h nonstop underwater endurance performance could lead to a generalized inflammatory response. The exceptionality of such a world record attempt, when compared to other shorter trials of the kind by Jerry Hall in 2013, 145 h, 31 min, 23 s at 3.6 m underwater, or Paolo De Vizzi in 2016, 51 h and 56 min at 10 m underwater, was determined by the equipment with which the athlete spent 36 h at 10 m underwater, consisting of a semi-dry suit, standard gas tank, and traditional second-stage diving regulator. Therefore, there was increased difficulty due to cold stress and contraction of facial muscles to hold the regulator, which forced sleep deprivation. The same athlete prior to this performance successfully realized 24-h-, 29-h-, and 32-h-long scuba diving. To our knowledge, there have been no other studies involving physiological responses caused by such long underwater performances, although there are scarce studies related to cold water survival in different types of clothing (Drapher and Hodgson 2008).

This extraordinary performance by an "elite sub" led us to investigate the pathophysiological effects of nonstop underwater endurance lasting for 36 h at 10 m (33 ft) depth, with 20 °C open water temperature. We assessed inflammatory markers, hormones, and a general hematological profile.

2 Methods

The "Endless Diving Project-Step 36" was performed in the Maratea harbor waters on September 12–13, 2014. The athlete was a 41-year-old man, expert diver having a 10-year streak of underwater performance (2004, 24 h; 2006, 29 h; and 2007, 32 h). The diver wore a 7-mm semi-dry suit with cuffs and collar of single-lined neoprene, a standard gas tank, and standard fittings for boots, gloves, and fins. The athlete ate food at his discretion every hour, following a standardized diary diet, with the help of the assisting divers, bringing hot enriched liquids and snacks, consisting mainly of cheese and chocolate, apples and bananas. The diver was breathing compressed ambient air, using a scuba diving gear without any changes in the gas mixture. The suit was periodically cleaned with argon insufflation to avoid skin irritation due to urine stagnation. He stayed on a platform submerged

10 m deep and made periodically brief rounds to avoid boredom and to counteract cold stress. Medical assistance was standing by, ready to counteract hypothermia, hypoglycemia, or any other complaints. Vital parameters such as heart rate (HR), blood pressure (BP), peripheral blood oxygen saturation (SpO_2), and body temperature and weight were assessed before and after the performance. Water temperature ranged from 16 to 25 °C. Venous blood samples were taken at baseline before the dive, immediately post-dive (at 22:00 pm), the day after (Day 1), 4 days after (Day 4), 12 days after (Day 12), and the 1 month (Month 1) and 41 months after the dive (Month 41). All the samples, except those taken immediately post-dive, were collected at 8:00 am. The samples were put on ice and transported without delay to the laboratory of the Casa Sollievo della Sofferenza Hospital in San Giovanni Rotondo in Italy, where they were centrifuged for 10 min at 3000 rpm and kept at −80 °C for further analysis.

We assessed the hematological and hormonal profiles in peripheral blood. The hematological indices were red blood cell (RBC), mean cell volume (MCV), red blood cell distribution width (RDW-CV), hematocrit (HCT), hemoglobin (HGB), mean corpuscular hemoglobin (MCH), mean corpuscular hemoglobin concentration (MCHC); white blood cell (WBC) count, neutrophils, lymphocytes, monocytes, eosinophils, basophils; and platelets (PLT), mean platelet volume (MPV), platelet distribution width (PDW), and the plateletcrit which is a measure of total platelet mass. The hormonal indices were follicle-stimulating hormone (FSH), luteinizing hormone (LH), prolactin, 17β-estradiol, testosterone; thyroid hormones, free thyroxine (T4), free triiodothyronine (T3), and thyroid stimulating hormone (TSH); and stress hormones, cortisol, adrenocorticotropic hormone (ACTH), insulin, somatotropin, and melatonin. In addition, we measured the content of monoamine neurotransmitters: serotonin, dopamine, adrenaline, and noradrenaline. Finally, we measured the following markers of inflammation: high sensitivity C-reactive protein (CRP), erythrocyte sedimentation rate (ESR), and the prostate-specific antigen (PSA).

3 Results

3.1 Vital Signs and Body Weight

There was a marked reduction in body weight after the 36-h dive, from 96.0 kg to 90.5 kg. BP and HR slightly decreased, from 130/75 to 120/68 mmHg and from 68 to 64 beats/min, respectively. The SpO_2 remained in a normal range of 98–100%, and body temperature slightly increased after from 36.5 °C to 36.9 °C.

3.2 Hematologic Profile

There were outstanding acute responses in the hematological profile after the dive. The WBC increased by 123% immediately post-dive; the increase reverted to the baseline level at Day 1 post-dive, with little fluctuations during the further follow-up. The PLT and RBC fluctuated around the baseline level throughout (Fig. 1). The percentage of neutrophils and monocytes increased by 230% and 55%, respectively, immediately post-dive, and eosinophils increased by 160% at Day 1 (Fig. 2). In contrast, there were

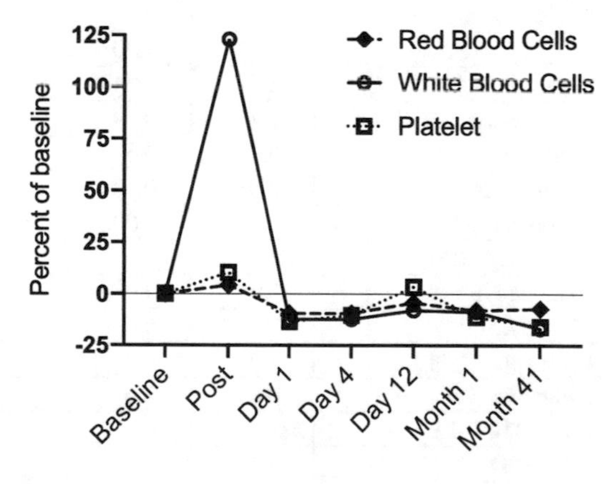

Fig. 1 Peripheral blood cell counts following the world record-breaking 36-h-long underwater endurance performance

reductions in basophils and lymphocytes by 75% and 41%, respectively, immediately post-dive (Fig. 2). Specific changes in hematologic indices in response to the 36-h dive are given in detail in Table 1.

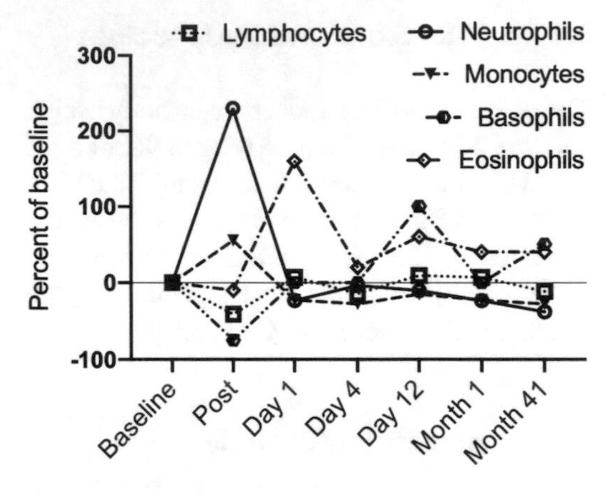

Fig. 2 White blood cell subpopulations following the world record-breaking 36-h-long underwater endurance performance

The HGB and HCT slightly increased immediately post-dive by about 6% and then decreased by 6% of the baseline level during Day1–Day4, gradually reverting throughout follow-up (Fig. 3a). There was an increase in CRP immediately post-dive, followed by a fall back to baseline and below during further follow-up. In contrast, ESR showed a sharp reduction post-dive, lasting for the first day, followed by a rebound to the pre-dive level, with some fluctuations up to Month 4 (Fig. 3b).

3.3 Monoamine Neurotransmitters and Hormones

Blood content of serotonin and dopamine outstandingly increased, and conversely that of adrenalin decreased, at Day 1 post-dive, gradually reverting to the baseline level throughout the long-term follow-up (Fig. 4a). Changes in

Table 1 Hematological profile following the world record-breaking 36-h-long underwater endurance performance

	Baseline	Immediately after	Day 1	Day 4	Day 12	Month 1	Month 41
RBC ($\times 10^6$/L)	5.70	5.94	5.20	5.16	5.46	5.25	5.29
MCV (fL)	83.0	85.3	83.8	84.2	84.5	84.0	85.5
RDW-CV (%)	14.3	16.4	13.2	13.6	13.7	14.4	13.5
HCT (%)	47.3	50.7	42.6	43.4	46.1	44.1	45.3
HGB (g/dL)	15.7	16.5	14.6	14.8	15.7	15.4	15.4
MCH (pg)	27.5	27.8	28.2	28.7	28.8	29.3	29.1
MCHC (g/dL)	33.2	32.6	34.6	34.1	34.1	34.9	34.0
WBC ($\times 10^3$/uL)	5.30	11.20	4.59	4.41	4.99	4.59	3.81
Neutrophils ($\times 10^3$/μL)	2.86	9.44	2.20	2.76	2.57	2.20	1.78
Lymphocytes ($\times 10^3$/μL)	1.7	0.99	1.82	1.46	1.86	1.82	1.51
Monocytes ($\times 10^3$/μL)	0.4	0.62	0.31	0.29	0.34	0.31	0.29
Eosinophils ($\times 10^3$/μL)	0.05	0.10	0.13	0.06	0.08	0.07	0.07
Basophils ($\times 10^3$/μL)	0.02	0.00	0.02	0.02	0.04	0.02	0.03
Neutrophils (%)	54.9	84.6	59.2	58.7	51.6	48.0	46.8
Lymphocytes (%)	33.8	8.9	30.1	31.0	37.3	39.7	39.6
Monocytes (%)	7.9	5.6	6.4	6.2	6.8	6.7	7.6
Eosinophils (%)	1.0	0.9	1.4	1.2	1.5	2.8	1.7
Basophils (%)	0.4	0.0	0.4	0.4	0.8	0.5	0.7
PLT ($\times 10^3$/μL)	232	256	201	208	240	206	195
MPV (fL)	9.5	8.8	7.1	7.2	6.7	7.7	9.0
PDW (%)	49.9	55.9	55.7	55.2	53.0	52.4	49.7
Plateletcrit (%)	0.22	0.13	0.14	0.15	0.16	0.16	0.18

RBC red blood cells, *RDW-CV* red blood cell distribution width, *HCT* hematocrit, *HGB* hemoglobin, *MCH* mean corpuscolar hemoglobin, *MCHC* mean corpuscolar hemoglobin concentration, *WBC* white blood cells, *PLT* platelets, *MPV* mean platelet volume, *PDW* platelet distribution width

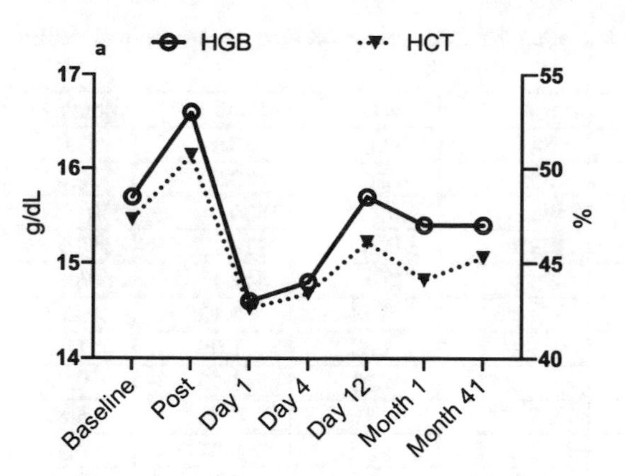

Fig. 3 Hematological and inflammatory profiles following the world record-breaking 36-h-long underwater endurance performance; (a) hemoglobin (HGB) and hematocrit (HCT) and (b) C-reactive protein (CRP) and erythrocyte sedimentation rate (ESR)

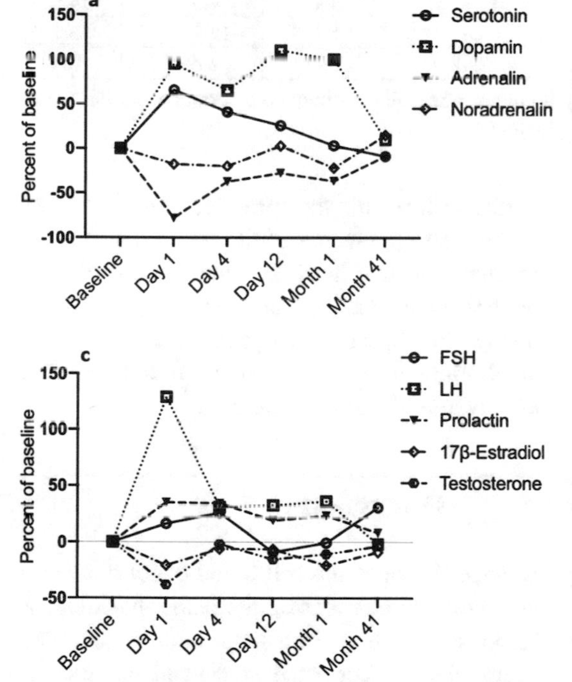

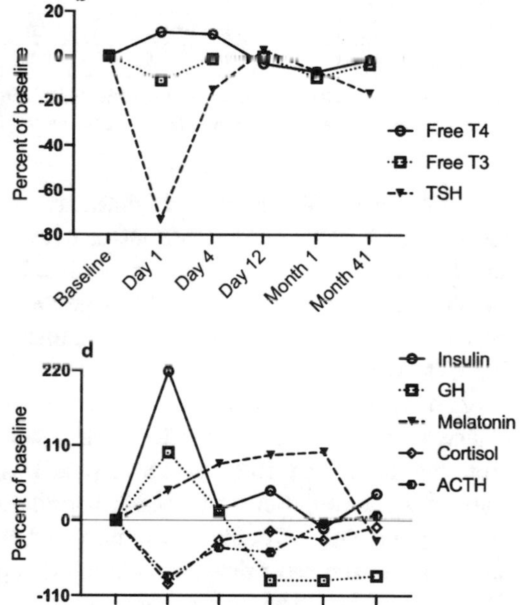

Fig. 4 Neurotransmitter and hormonal changes following the world record-breaking 36-h-long underwater endurance performance; (a) monoamine neurotransmitters, (b) thyroid hormones, (c) sex hormones, and (d) stress and anabolic indices

noradrenalin were unremarkably as its content fluctuated about the baseline level. The ratio of serotonin (pg/mL) to dopamine (pg/mL) amounted to 2900 before the dive, and it decreased at the successive time points post-dive as follows: 2462 (Day 1), 2485 (Day 4), 1738 (Day 12), 1500 (Month 1), and rebounding in the end closely to the baseline level with the figure of 2409 at Month 41.

Table 2 Monoamines and hormones in peripheral blood following the world record-breaking 36-h-long underwater endurance performance

	Before	Day 1	Day 4	Day 12	Month 1	Month 41
Serotonin (μg/L)	58.0	96.0	82.0	73.0	60.0	53.0
Dopamine (pg/mL)	20.0	39.0	33.0	42.0	40.0	22.0
Serotonin/dopamine ratio	2900	2462	2485	1738	1500	2409
Adrenaline (pg/mL)	173.6	37.7	108.7	125.5	110.1	158.5
Noradrenaline (pg/mL)	527.4	434.4	423.1	543.1	412.4	606.6
Free T3 (pg/mL)	3.45	3.07	3.40	3.44	3.12	3.32
Free T4 (ng/DL)	1.13	1.25	1.24	1.09	1.05	1.11
TSH (μUI/mL)	1.18	0.32	1.00	1.21	1.10	0.98
FSH (mUI/mL)	4.51	5.23	5.67	4.10	4.48	5.89
LH (mUI/mL)	4.46	10.21	5.84	5.92	6.07	4.37
Testosterone (ng/mL)	5.46	3.39	5.35	6.30	4.87	5.25
Cortisol (μg/dL)	12.40	0.90	8.76	10.47	8.87	11.20
Prolactin (ng/mL)	5.18	7.00	6.96	6.16	6.38	5.59
17β-Estradiol (pg/mL)	47.24	37.47	44.26	44.20	37.43	42.92
GH (ng/mL)	0.80	1.60	0.90	0.10	0.10	0.15
Insulin (μUI/mL)	10.5	33.5	12.2	15.1	9.3	14.6
ACTH (pg/mL)	14.6	2.6	8.8	7.8	14.1	15.7
Melatonin (pg/mL)	1.32	1.90	2.42	2.59	2.65	0.92

T3 triiodothyronine, *T4* thyroxine, *TSH* thyroid stimulating hormone, *FSH* follicle-stimulating hormone, *LH* luteinizing hormone, *GH* growth hormone, *ACTH* adrenocorticotropic hormone

Blood content of TSH showed a drastic reduction by about 73% at Day 1 post-dive, with an equally drastic recovery at Day 12 post-dive. On the other hand, changes in T3 and T4 hormones in response to the 36-h dive turned out unremarkable (Fig. 4b).

Referring to sex hormones, the dive produced an increase in blood LH and a reduction in testosterone by 129% and 38%, respectively, at Day 1 post-dive, with the content of both hormones reverting gradually and fully toward the pre-dive values at the successive time points of follow-up till Month 41. The other sex hormones investigated changed less. Prolactin and FSH increased by 35% and 16%, respectively; 17-β-estradiol decreased by about 21% of baseline, all with partial recovery up to Month 1 (Fig. 4c).

There were important changes noticed in the remaining hormones investigated in response to the dive. Insulin increased at Day 1 by 219%, reverting to the pre-dive value at Day 4. Melatonin was gradually increasing up to Day 4 reaching 83%; the increase persisted up to Month 1. Conversely, ACTH and cortisol were reduced at Day 1 post-dive by 82% and 93%, respectively, with a partial return to the pre-dive level at Day 4. Somatotropin increased at Day 1 only to return to the pre-dive level at Day 4, but then it underwent a sustained drop level up to Month 4 (Fig. 4d). Specific changes in neurotransmitter and hormonal values in response to the 36-h dive are given in detail in Table 2.

4 Discussion

Changes in the white cell blood count draw special attention considering the pathophysiological responses to this 36-h-long underwater dive. These changes consisted of evident increases in peripheral blood neutrophils, monocytes, and eosinophils and a reduction of basophils and lymphocytes. Neutrophils are essential blood cells that reflect systemic stress and inflammation. Inflammatory markers, such as C-reactive protein, stimulate granulopoiesis and rapid release of neutrophils from the bone marrow (Summers et al. 2010). The abundance of neutrophils, coupled with a brief, 6–8 h long, circulating half-life, may explain their drastic reduction noticed the day after. An acute increase in neutrophils, with the concurrent lymphocytopenia, has been suggested to reflect the effect of prolonged physical stress, such as that found in the ultra-marathon run in a study of Shin and Lee (2013).

Eosinophils are granular leukocytes implicated in the pathogenesis of a range of disorders. An increase in eosinophils in peripheral blood has been linked to vasculitis (Khoury et al. 2014) Thus, we could hypothesize that the 36-h dive generates a hematological and clinical conditions resembling a primary vasculitis syndrome, a type of vasculitis caused by inflammation of blood vessel walls (Okazaki et al. 2017). The plausible mechanisms, in case of the prolonged nonstop dive, might be cold stress, dysfunction of endothelial cells, physical immobility, cognitive dysfunction, sleep deprivation, and a fixed spatial body position. The hypothesis of an eosinophil vasculitis-like syndrome finds support in the simultaneous increase of C-reactive protein with accompanying leukocytosis in response to the prolonged dive of this study. A drop in ESR, somehow opposing the inflammatory response, is explicable by increased hematocrit as ESR is inversely related to hematocrit (Allen 1988). Perovic et al. (2017) have demonstrated an increase of neutrophils without changes in hematocrit after a short-term diving exposure. Thus, increase in hematocrit in the present prolonged drive, as well as in longer dives reported in other studies (Castagna et al. 2015), could be driven by dehydration, consequent to fluid loss and a lack of possibility of its timely replenishment. Although we did not measure body fluids, the marked reduction of body weight after dive strongly supports the suggestion of dehydration. Hyperbaric condition, in itself, facilitates dehydration of the diver's body (Kaczerska et al. 2019).

Monoamine neurotransmitters are essential for maintaining central nervous system activity (Wu et al. 2016; Hussain and Lokhandwala 2003; Timmerman et al. 1999). Recently, central origin of fatigue draws increasing attention, although fatigue also has peripheral causes such as glycogen depletion or cardiovascular strain. Brain serotonin and dopamine are related to the perception of fatigue, a feeling that reduces the intensity of physical exercise during prolonged strenuous exercise (Meeusen and Roelands 2010; Meeusen et al. 2006; Newsholme and Blomstrand 2006). Thus, these neurotransmitters control the exercise performance (Cordeiro et al. 2017). Davis and Bailey (1997) have shown that an increase or decrease in brain serotonin activity, which may occur during prolonged exercise, hastens or delays fatigue, respectively. Those authors also highlight the regulatory role of serotonin/dopamine interaction for exercise performance. A low ratio of these neurotransmitters would facilitate better performance, and a high ratio would rather decrease motivation and could lead to lethargy. In congruity with those findings, we showed in the present study that the serotonin/dopamine ratio decreased during the prolonged 36-h dive, which may be an expression of brain activation helping in the record achievement. Considering dopamine and serotonin separately, the former remained enhanced up to 1 month after the performance, while the latter gradually decreased down to the baseline level. We may speculate that these changes reflect the maintaining of a high level of positive mood related to serotonin, and rewarding related to dopamine, for the weeks to come in the wake of a record-breaking performance. Finally, as serotonin, dopamine, and noradrenaline are also implicated in the control of thermoregulation, a shift in the content of these neurotransmitters may contribute to the thermal and fatigue resistance (Bridge et al. 2003). Anegg et al. (2002) have reported an increment of adrenalin and noradrenalin after short-term immersions. In contrast, in this study we noticed a decrease in both catecholamines after the prolonged immersion. This decrease may lead to a reduction of HR and BP after the prolonged dive, the effect exerted via the autonomic nervous system, which would be opposite to that observed in short-term physical exercise or diving. Considering that sympathetic reactivity is reduced after psychological exhaustion, we may also hypothesize that the present extreme performance produced a psychological stress which impaired the otherwise augmented response of the autonomic nervous system to short-term efforts.

An increase in serum prolactin we noticed in this study post-diving, along with a reduction in 17β-estradiol and testosterone, is congruous with the findings reported in short-term scuba diving in males (Verratti et al. 2019). Prolactin is known to reflect the activation of physical and mental vigilance (Anegg et al. 2002), a likely feature of the world record-breaking performance. A drastic reduction in testosterone, on the other side, seems to reflect an acute primary testicular dysfunction, which could lead to a compensatory increase in luteinizing hormone. A negative feedback between testosterone and luteinizing hormone has a role in the formation of spermatozoa to maintain the fertility status is known (Foresta et al. 1997).

Referring to metabolism and hormones, the present findings on the effects of a prolonged immersion show specific changes the day after consisting of a distinct increase in insulin, a moderate increase in growth hormone, and a reduction of thyroid stimulating hormone, the changes that rather promptly reverted to the baseline level 4 days post-dive. A decrease in cortisol we noticed could be due to the action of psychophysical stress, including also sleep deprivation that is known to increase the brain content of melatonin (Salín-Pascual et al. 1988). Marlinge et al. (2019) have reported a reduction in serum cortisol after scuba diving, a likely effect of hypoxia (Wright et al. 2015).

5 Conclusions

The "Endless Diving Project-Step 36" provided a unique opportunity to study inflammatory, hematological, hormonal, and neurotransmitter responses to an extremely prolonged nonstop underwater performance. We found a propensity for the development of acute inflammatory state, likely with dehydration, accompanied by a vigilant mood condition with features of long-term rewarding, despite the apparent exhaustion. The pathophysiological changes evoked by prolonged diving resembled in some respects those that are caused by acute diving exposure or long-term physical exercise. There was a rather prompt return to normal homeostatic control in about 1–4 days after the performance. Further studies are needed to clarify the effect of circadian misalignment, caused mainly by sleep deprivation, on the cytokine and hormonal profile. A full understanding of human pathophysiology requires the knowledge gained in studies on homeostatic control mechanisms during extreme body conditions.

Acknowledgments Our special thanks go to the athlete Francesco Colletta (www.endlessdiving.com). We also thank the nutritionist, divers, and medical staff who assisted the athlete during this record-breaking dive, as well as representatives of many a firm who took care of technical and logistical procedures: Vittorio Bianchini (NASE® Italian Dept.); Lazzaro Di Mauro, director of Clinical Analysis Laboratory; and Gaetano Desina, manager of Clinical Laboratory (Ospedale San Giovanni Rotondo FG) in Italy. This work was supported by a grant from the Department of Psychological, Health and Territorial Sciences, "G. d'Annunzio" University of Chieti-Pescara, Italy.

Conflicts of Interest The authors declare no conflicts of interest related to this article.

Ethical Approval All procedures performed in studies involving human participants were in accordance with the ethical standards of the institutional and/or national research committee and with the 1964 Helsinki Declaration and its later amendments or comparable ethical standards. Study protocol was approved by the Ethics Committee of "G. D'Annunzio" University of Chieti-Pescara, Italy.

Informed Consent Written informed consent was obtained from the participant of this study.

References

Allen BV (1988) Relationships between the erythrocyte sedimentation rate, plasma proteins and viscosity, and leucocyte counts in thoroughbred racehorses. Vet Rec 122:329–332

Anegg U, Dietmaier G, Maier A, Tomaselli F, Gabor S, Kallus KW, Smolle-Jüttner FM (2002) Stress-induced hormonal and mood responses in scuba divers: a field study. Life Sci 70:2721–2734

Bridge MW, Weller AS, Rayson M, Jones DA (2003) Responses to exercise in the heat related to measures of hypothalamic serotonergic and dopaminergic function. Eur J Appl Physiol 89:451–459

Castagna O, Desruelle AV, Blatteau JE, Schmid B, Dumoulin G, Regnard J (2015) Alterations in body fluid balance during fin swimming in 29 °C water in a population of special forces divers. Int J Sports Med 36:1125–1133

Catchpole HR, Gersh I (1947) Pathogenetic factors and pathological consequences of decompression sickness. Physiol Rev 27:360–397

Cordeiro LMS, Rabelo PCR, Moraes MM, Teixeira-Coelho F, Coimbra CC, Wanner SP, Soares DD (2017) Physical exercise-induced fatigue: the role of serotonergic and dopaminergic systems. Braz J Med Biol Res 50:e6432

Davis JM, Bailey SP (1997) Possible mechanisms of central nervous system fatigue during exercise. Med Sci Sports Exerc 29:45–57

Drapher N, Hodgson C (2008) Adventure sport physiology. Wiley-Blackwell, Hoboken. ISBN:978-0-470-01510-0

Eftedal I, Ljubkovic M, Flatberg A, Jørgensen A, Brubakk AO, Dujic Z (2013) Acute and potentially persistent effects of scuba diving on the blood transcriptome of experienced divers. Physiol Genomics 45:965–972

Foresta C, Bordon P, Rossato M, Mioni R, Veldhuis JD (1997) Specific linkages among luteinizing hormone, follicle-stimulating hormone, and testosterone release in the peripheral blood and human spermatic vein: evidence for both positive (feed-forward) and negative (feedback) within-axis regulation. J Clin Endocrinol Metab 82:3040–3046

Grivennikov SI, Greten FR, Karin M (2010) Immunity, inflammation, and cancer. Cell 140:883–899

Hussain T, Lokhandwala MF (2003) Renal dopamine receptors and hypertension. Exp Biol Med (Maywood) 228(2):134–142

Kaczerska D, Siermontowski P, Kozakiewicz M, Krefft K, Olszański R (2019) Dehydration of a diver during a hyperbaric chamber exposure with oxygen decompression. Undersea Hyperb Med 46(2):185–188

Khoury P, Grayson PC, Klion AD (2014) Eosinophils in vasculitis: characteristics and roles in pathogenesis. Nat Rev Rheumatol 10:474–483

Marlinge M, Coulange M, Fitzpatrick RC, Delacroix R, Gabarre A, Lainé N, Cautela J, Louge P, Boussuges A, Rostain J-C, Guieu R, Joulia FC (2019) Physiological stress markers during breath-hold diving and SCUBA diving. Physiol Rep 7:e14033

Meeusen R, Roelands B (2010) Central fatigue and neurotransmitters, can thermoregulation be manipulated? Scand J Med Sci Sports 20(Suppl 3):19–28

Meeusen R, Watson P, Hasegawa H, Roelands B, Piacentini MF (2006) Central fatigue: the serotonin hypothesis and beyond. Sports Med 36(10):881–909

Newsholme EA, Blomstrand E (2006) Branched-chain amino acids and central fatigue. J Nutr 136 (1 Suppl):274S–276S

Okazaki T, Shinagawa S, Mikage H (2017) Vasculitis syndrome-diagnosis and therapy. J Gen Fam Med 18 (2):72–78

Peng Z, Ren P, Kang Z, Du J, Lian Q, Liu Y, Zhang JH, Sun X (2008) Up-regulated HIF-1alpha is involved in the hypoxic tolerance induced by hyperbaric oxygen preconditioning. Brain Res 1212:71–78

Perovic A, Nikolac N, Braticevic MN, Milcic A, Sobocanec S, Balog T, Dabelic S, Dumic J (2017) Does recreational scuba diving have clinically significant effect on routine haematological parameters? Biochem Med (Zagreb) 27(2):325–331

Salín-Pascual RJ, Ortega-Soto H, Huerto-Delgadillo L, Camacho-Arroyo I, Roldán-Roldán G, Tamarkin L (1988) The effect of total sleep deprivation on plasma melatonin and cortisol in healthy human volunteers. Sleep 11:362–369

Shin YO, Lee JB (2013) Leukocyte chemotactic cytokine and leukocyte subset responses during ultra-marathon running. Cytokine 61:364–369

Sparmann A, Bar-Sagi D (2004) Ras-induced interleukin-8 expression plays a critical role in tumor growth and angiogenesis. Cancer Cell 6:447–458

Summers C, Rankin SM, Condliffe AM, Singh N, Peters AM, Chilvers ER (2010) Neutrophil kinetics in health and disease. Trends Immunol 31:318–324

Thom SR, Yang M, Bhopale VM, Huang S, Milovanova TN (2011) Microparticles initiate decompression-induced neutrophil activation and subsequent vascular injuries. J Appl Physiol 110:340–351. (1985b)

Thom SR, Milovanova TN, Bogush M, Yang M, Bhopale VM, Pollock NW, Ljubkovic M, Denoble P, Madden D, Lozo M, Dujic Z (2013) Bubbles, microparticles, and neutrophil activation: changes with exercise level and breathing gas during open-water SCUBA diving. J Appl Physiol 114:1396–1405. (1985a)

Timmerman W, Cisci G, Nap A, de Vries JB, Westerink BH (1999) Effects of handling on extracellular levels of glutamate and other amino acids in various areas of the brain measured by microdialysis. Brain Res 833:150–160

Verratti V, Bondi D, Jandova T, Camporesi E, Paoli A, Bosco G (2019) Sex hormones response to physical hyperoxic and hyperbaric stress in male scuba divers: a pilot study. Adv Exp Med Biol 1176:53–62

Wilmshurst P (1998) Diving and oxygen. BMJ 317 (7164):996–999

Wright KP, Drake AL, Frey DJ, Fleshner M, Desouza CA, Gronfier C, Czeisler CA (2015) Influence of sleep deprivation and circadian misalignment on cortisol, inflammatory markers, and cytokine balance. Brain Behav Immun 47:24–34

Wu D, Xie H, Lu H, Li W, Zhang Q (2016) Sensitive determination of norepinephrine, epinephrine, dopamine and 5-hydroxytryptamine by coupling HPLC with [Ag(HIO6)2](5-) -luminol chemiluminescence detection. Biomed Chromatogr 30(9):1458–1466

Yang M, Milovanova TN, Bogush M, Uzun G, Bhopale VM, Thom SR (2012) Microparticle enlargement and altered surface proteins after air decompression are associated with inflammatory vascular injuries. J Appl Physiol 112:204–211 z. (1985)

Yoshida N, Ikemoto S, Narita K, Sugimura K, Wada S, Yasumoto R, Kishimoto T, Nakatani T (2002) Interleukin-6, tumour necrosis factor alpha and interleukin-1beta in patients with renal cell carcinoma. Br J Cancer 86:1396–1400

Adv Exp Med Biol - Clinical and Experimental Biomedicine (2021) 11: 89–97
https://doi.org/10.1007/5584_2020_554

Published online: 25 June 2020

Respiratory Muscle Strength and Ventilatory Function Outcome: Differences Between Trained Athletes and Healthy Untrained Persons

Marina O. Segizbaeva and Nina P. Aleksandrova

Abstract

It is known that the maximum mouth inspiratory pressure (MIP) and expiratory pressure (MEP) vary with age, weight, height, and skeletal muscle mass. However, the influence of physical training on ventilatory function outcomes is an area of limited understanding. The aim of this study was to investigate the respiratory muscle strength and its relation to spirometry variables in untrained healthy persons versus trained athletes. MIP and MEP were assessed in 22 power athletes and 28 endurance athletes, and in 24 age- and sex-matched normal healthy subjects (control group). The measurement was done with a mouth pressure meter. We found that respiratory muscle strength and ventilatory function in endurance athletes were outstandingly superior to that in power athletes; the latter's muscle strength was better than that of healthy untrained controls. Both MIP and MEP significantly correlated with the maximum voluntary ventilation (MVV) in both power athletes and controls, but not so in endurance athletes. The corollary is that the intensive endurance training could result in the improvement of respiratory muscle strength, meeting the maximum upper limit of functional reserve of respiratory muscles and the corresponding ventilation. On the other hand, targeted training of respiratory muscle strength may be an effective strategy to increase ventilatory function in power athletes, particularly those having a low maximum inspiratory and expiratory pressure, and in less physically fit healthy persons.

Keywords

Athletes · Expiratory pressure · Inspiratory pressure · Maximum voluntary ventilation · Spirometry · Respiratory muscle strength

1 Introduction

The primary function of the respiratory muscular or motor system is to maintain alveolar ventilation according to the metabolic needs. Respiratory muscle function is determined by the strength and endurance of respiratory muscles (ATS/ERS 2002). The unique structural characteristics of respiratory muscles combined with neural regulation of breathing indicate that the capacity of these muscles for pressure generation usually exceeds the demands placed on them (Romer and Polkey 2008). However, there is sufficient evidence that inspiratory muscles' ability to generate force may

M. O. Segizbaeva (✉) and N. P. Aleksandrova
Laboratory of Respiration Physiology, I.P. Pavlov Institute of Physiology RAS, St. Petersburg, Russia
e-mail: segizbaevamo@infran.ru

be reduced during heavy exercise, maintained for a prolonged time. Several studies confirm that the global performance may be limited by respiratory muscle fatigue in both trained and untrained subjects (Segizbaeva et al. 2013; Wells and Norris 2009; Johnson et al. 1996). High-intensive exercise triggers respiratory muscle metaboreflex that causes peripheral vasoconstriction, limits blood flow to working lower limb muscles, and accelerates muscle fatigue (Janssens et al. 2013; Romer and Dempsey 2006). The metaboreflex contributes to the exercise limitation in healthy subjects during heavy endurance exercise (Wüthrich et al. 2013). In professional athletes with higher respiratory muscle strength, triggering of the respiratory muscle metaboreflex is delayed, making them achieve better sportive results (Witt et al. 2007). It is well known that respiratory muscles can be trained like other skeletal muscles in healthy subjects, athletes, and in patients with dyspnea (Sales et al. 2016; Segizbaeva et al. 2015; Neves et al. 2014). Respiratory muscles adapt to specific respiratory muscle training and to whole body exercise training in relation to ventilatory function, respiratory muscle endurance, and maximum strength (Boutellier et al. 1992). The function of respiratory muscles, like other skeletal muscles, improves in response to regular training. Exercise capacity and respiratory muscle strength are important to athletes, especially competitive athletes (HajGhanbari et al. 2013; Illi et al. 2012). It is established that participation in sports is associated with respiratory adaptation whose extend depends on the type of activity (Lazovic et al. 2015).

Spirometry is the gold standard for the assessment of ventilatory function (Miller et al. 2005). Activation and powerful contractions of different groups of inspiratory muscles during the forced vital capacity maneuver and expiratory muscles during forced expiration provide rapid changes in intra-thoracic pressure, which ensures the actual fulfillment of airflow (Tiller and Simpson 2018). Both primary and all accessory respiratory muscles are actively involved in forced inhalation and exhalation maneuvers during testing. However, the influence of respiratory muscle strength on ventilatory function outcome has not yet been investigated. Therefore, we addressed the issue in this study by investigating the maximum inspiratory and expiratory pressures (MIP and MEP, respectively) and the relationship between respiratory muscle strength and the results of spirometry variables in healthy untrained persons and trained athletes.

2 Methods

2.1 Study Subjects

In total, there were 74 young healthy, non-smoking male subjects enrolled into the study. They were divided into three groups, based on preferable sport activity. The first two groups were athletes: Group I – 28 endurance athletes (6 middle- and long-distance runners, 14 swimmers, 4 skiers, and 4 multi-event athletes (pentathlon and triathlon)), Group II – 22 power athletes (7 weightlifters, 1 arm-wrestlers, 3 boxers, and 11 sambo, judo, taekwondo, and karate athletes). All of them were postgraduate students of the fourth semester in the Military Sport Institute in St. Petersburg, Russia. All athletes had a sports category from 1 to Sport Master and had no less than 18 h of training *per* week. Group III was the control group, consisting of 24 age- and sex-matched subjects, who were physically active having some regular amateur exercise but were not engaged in any specialist athletic training. All the participants underwent physical examination before the enrollment. They were free of any overt cardiovascular, ventilatory, neurological, and any other disorders. The subjects were instructed not to consume alcohol, caffeine beverages, or to use any inhalers or drugs 12 h preceding the tests. They were also advised to maintain adequate hydration and have a light meal up to 2 h before the tests. The anthropometric measurements included body height and body mass. Body mass index (BMI) was calculated by dividing body mass in kilograms by height in meters squared.

2.2 Maximum Inspiratory (MIP) and Expiratory (MEP) Pressure

Respiratory muscle strength was assessed by measuring MIP and MEP. Both were measured using a portable mouth meter (FusionCare Inc., San Diego, CA), according to the ATS/ERS (2002) standards. A disposable antibacterial filter and a latex mouthpiece were used for each subject. The procedure was a priori explained and demonstrated by a researcher. Subjects performed the test in the standing position with a nose clip in place, with the meter held in a hand. MIP and MEP were recorded at the mouth during a quasi-short maximum inspiration or expiration against occluded airways, respectively. To measure MIP, subjects were instructed to breath out to residual volume (RV) and then to perform maximum inspiratory effort as hard and quickly as possible, and to sustain this inspiration for 3–4 s (Troosters et al. 2005). To measure MEP, they were asked to breath in to total lung capacity (TLC) and then expire as quickly as possible for 3–4 s. During the tests, the subjects were verbally encouraged to achieve a maximum result. To prevent closure of the glottis and to avoid significant pressure generation by the cheek muscles, a small leak was introduced in the measuring device. The subjects also pressed the palm of a hand against the cheeks during testing. For both maneuvers, measurements were repeated at 1 min intervals to prevent the development respiratory muscle fatigue. Data were reported in cmH_2O. For the sake of convenience, MIP and MEP were expressed as the positive values. Each subject was tested at least five times, and the highest value that had less than 10% inter-measurement variability was taken for the analysis. All the tests were performed in the laboratory setting with the same instruments and techniques and by the same researcher.

2.3 Ventilatory Function

Ventilatory function tests were performed using spirometry, according to the ATS/ERS (2002) recommendations. Measurements were carried out in the laboratory setting between 9 and 12 am at room temperature 20–24 °C, atmospheric pressure 760 mmHg, and a relative humidity of 40–60%. All measurements were performed using a portable, automated MicroLoop USB spirometer equipped with disposable antibacterial filters (FusionCare Inc., San Diego, CA). Subjects performed the test in the sitting position with a nose clip in place and the spirometer device held in a hand. After the maximum inhalation, the subject was instructed to exhale as hard and as fast as possible, being encouraged to continue exhaling for at least 4 s, so that FEV_1 could be obtained. Tests were repeated a minimum of three times or until the two highest values were recorded, with 3–4 min intervals between trials. Besides FEV_1, the measurements included forced vital capacity (FVC), peak inspiratory and expiratory flows (PIF and PEF), and maximum voluntary ventilation (MVV). During the MVV test, subjects were asked to breathe deeply, quickly, and strongly for 12 s. The ventilation value recorded in 12 s was extrapolated to 1 min. Testing was carried out until three acceptable and two reproducible ventilation curves were obtained, not exceeding four attempts.

2.4 Statistical Analysis

Descriptive data were expressed as means ±SD and min–max ranges. Inter-group differences in each variable were assessed with a two-sample *t*-test. Pearson's correlation coefficient (*r*) was used to determine the strength of a relationship between MIP or MEP and the other ventilatory indices recorded. A p-value <0.05 defined statistically significant differences. Data were analyzed using the Microsoft Excel statistical package.

3 Results

The baseline anthropometric characteristics of the subjects are summarized in Table 1. There were not significant differences in the age, body height and mass, and BMI between the athletes and controls, albeit power athletes tended to have a higher BMI than the other two groups.

Table 1 Anthropometric characteristics in study groups

Variables	Athletes Endurance (*n* = 28)	Athletes Power (*n* = 22)	Controls (*n* = 24)
Age, years (range)	19.1 ± 0.4 (18–20)	19.4 ± 0.5 (18–20)	19.3 ± 0.6 (18–20)
BM, kg (range)	73.5 ± 6.6 (59–90)	76.6 ± 9.1 (65–103)	72.11 ± 9.3 (55–102)
BH, cm (range)	179.3 ± 6.0 (163–194)	176.8 ± 6.7 (165–187)	178.8 ± 6.7 (165–194)
BMI, kg/m^2 (range)	22.8 ± 1.2 (20.2–25.9)	24.5 ± 2.4 (20.0–32.5)	22.6 ± 2.6 (19.2–28.1)

Values are means ±SD
BM body mass, *BH* body height, *BMI* body mass index

Table 2 Maximum inspiratory (MIP) and expiratory pressure (MEP) and ventilatory function variables in the study groups

Variables	Athletes Endurance (*n* = 28)	Athletes Power (*n* = 22)	Controls (*n* = 24)
MIP, cmH_2O	162.60 ± 16.34**‡ (139–216)	150.50 ± 36.84** (90–218)	115.29 ± 24.98 (55–149)
MEP, cmH_2O	187.30 ± 31.40**‡‡ (149–275)	153.10 ± 49.38** (74–251)	118.33 ± 23.50 (68–157)
FVC, L	5.63 ± 0.65* (4.76–7.40)	5.20 ± 0.79 (3.65–6.69)	4.90 ± 0.73 (3.53–6.14)
FEV_1, L	5.10 ± 0.49* (4.15–6.31)	4.58 ± 0.52 (3.57–5.55)	4.55 ± 0.72 (2.60–5.98)
PIF, L/s	7.70 ± 1.24* (5.97–10.14)	6.97 ± 1.28 (4.25–8.67)	6.83 ± 1.61 (3.24–9.03)
PEF, L/s	10.50 ± 0.95* (7.85–11.20)	8.92 ± 1.27 (6.84–11.44)	8.70 ± 1.57 (6.09–10.90)
MVV, L/min	178.64 ± 16.65*‡ (141–225)	162.03 ± 19.22* (119–187)	148.66 ± 18.95 (99–176)

Values are means ±SD
FVC forced vital capacity, *FEV_1* forced inspiratory volume in one second, *PIF* peak inspiratory flow, *PEF* peak expiratory flow, *MVV* maximum voluntary ventilation
*$p < 0.05$, **$p < 0.01$ compared with control; ‡$p < 0.05$, ‡‡$p < 0.01$ compared with power athletes

MIP and MEP were significantly greater in both groups of athletes when compared to controls. However, the highest MIP and MEP values were recorded in endurance athletes, and these values were significantly greater than those in power athletes. MEP was higher than MIP in all the groups. Endurance athletes had a 41% and 58% higher MIP and MEP, while power athletes had a 30% and 29.4% larger MEP, respectively, compared to controls (Table 2).

The results of ventilatory function tests are shown in Table 2. The mean values of FVC, FEV_1, PIF, PEF, and MVV were significantly greater in endurance athletes when compared to controls. Ventilatory indices also tended to be greater in endurance than power athletes, with MVV reaching a significant difference. MVV also was the only variable in power athletes, which was greater than that in controls. The mean values of MVV were approximately 20% and 9% higher in endurance and power athletes than those in control subjects, respectively ($p < 0.05$).

We also found that MIP and MEP showed positive, albeit generally weak, correlations with

Table 3 Relationships between maximum inspiratory (MIP) and expiratory pressures (MEP) and ventilatory function variables in the study groups; Pearson's correlation coefficient *r*

Variables	FVC	FEV_1	PIF	PEF	MVV
Endurance athletes ($n = 28$)					
MIP	0.15 ns	0.10 ns	0.42*	0.16 ns	0.26 ns
MEP	0.20 ns	0.19 ns	0.16 ns	0.29 ns	0.10 ns
Power athletes ($n = 22$)					
MIP	0.40 ns	0.31 ns	0.27 ns	0.35 NS	0.58**
MEP	0.49*	0.43*	0.39	0.42*	0.59**
Controls ($n = 24$)					
MIP	0.35 ns	0.32 ns	0.29 ns	0.40 ns	0.64**
MEP	0.46*	0.52**	0.39	0.61**	0.61**

FVC forced vital capacity, FEV_1 forced inspiratory volume in one second, *PIF* peak inspiratory flow, *PEF* peak expiratory flow, *MVV* maximum voluntary ventilation, Pearson's correlation coefficient (*r*)
*$p < 0.05$; **$p < 0.01$; *ns,* non-significant

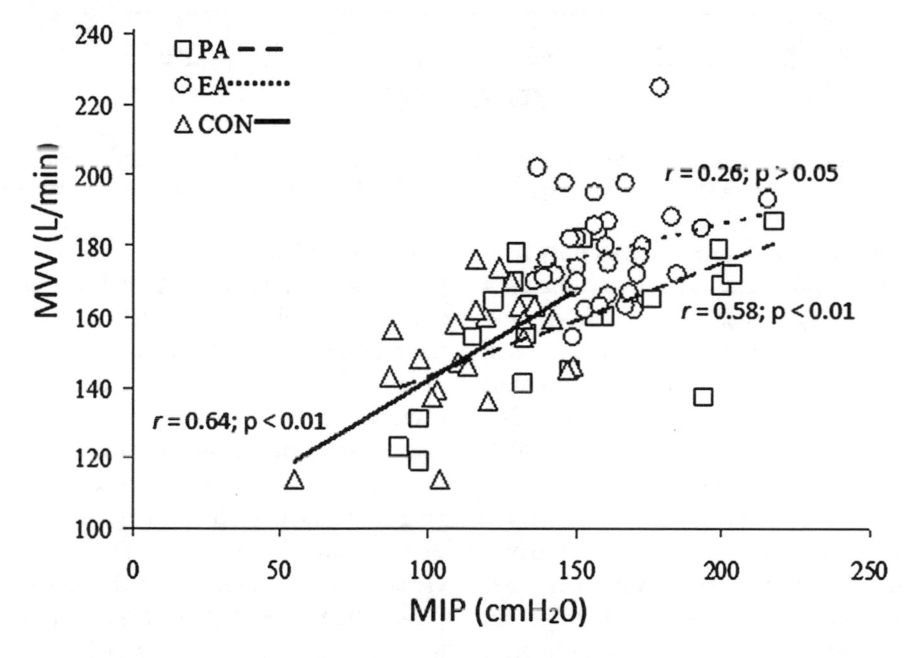

Fig. 1 Correlation between maximum inspiratory pressure (MIP) and maximum voluntary ventilation (MVV). *PA,* power athletes; *EA,* endurance athletes; *CON,* control untrained subjects. Lines are the least squares regression lines

the ventilatory variables (Table 3). In endurance athletes, the only significant correlation was that between MIP and PIF. In power athletes, akin to controls, MIP significantly associated with MVV only ($r = 0.58$ and $r = 0.64$, respectively, $p < 0.01$), while this relation in endurance athletes was weak and insignificant ($r = 0.26$; $p > 0.05$) (Fig. 1). On the other hand, MEP significantly correlated with all the variables, except for PIF, in both power athletes and controls. Again, the correlation of MEP with MVV was strong and significant in both controls and power athletes ($r = 0.61$ and $r = 0.59$, respectively, $p < 0.01$) (Fig. 2) but not so in endurance athletes.

4 Discussion

This study investigated the plausible connection between maximum inspiratory/expiratory

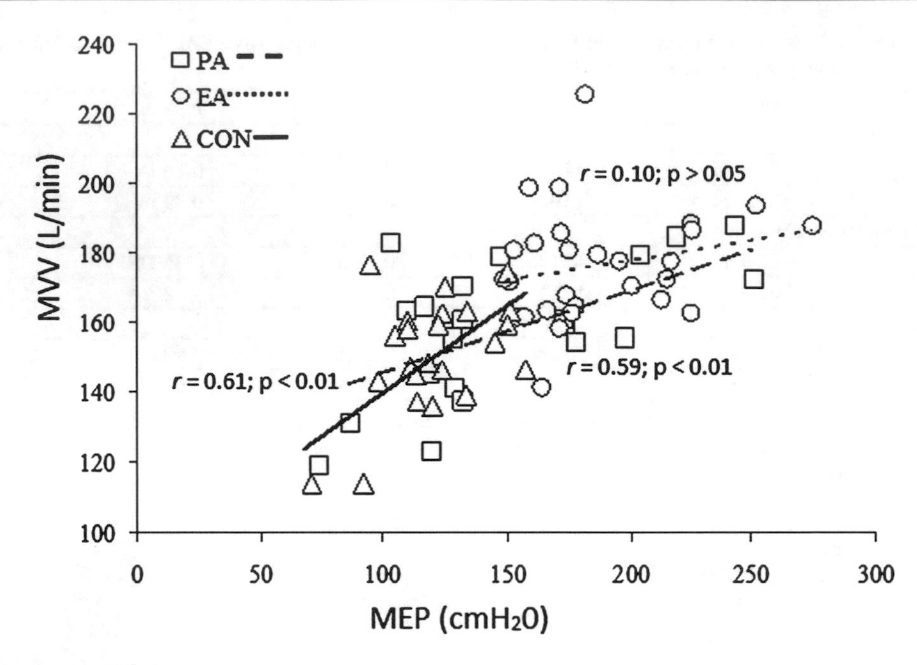

Fig. 2 Correlation between maximum expiratory pressure (MEP) and maximum voluntary ventilation (MVV). *PA*, power athletes; *EA*, endurance athletes; *CON*, control untrained subjects. Lines are the least squares regression lines

pressures, taken as surrogates of respiratory muscle strength, and basic ventilatory function variables in highly trained endurance and power athletes versus age-matched untrained male subjects. The main finding was that respiratory muscle strength and ventilatory function in endurance athletes were superior to those present in both power athletes and control subjects. Further, power athletes also had significantly higher levels of MIP and MEP, accompanied by increased maximum voluntary ventilation, when compared to control untrained subjects, although the differences were less expressive. These findings were in line with previous reports, which have demonstrated that respiratory muscle strength, as well as static and dynamic lung volumes are higher in swimmers (Rosser-Stanford et al. 2019; Lazovic-Popovic et al. 2016; Rong et al. 2008; Doherty and Dimitriou 1997; Armour et al. 1993; Clanton et al. 1987), soccer players (Ozmen et al. 2017), or rowers (Klusiewicz et al. 2008). Endurance athletes have a higher and prolonged requirement for gas exchange and thus also higher lung ventilation during exercise than power athletes and non-athletes. Part of an enhancing effect of physical training on ventilatory function might be attributable to higher respiratory muscle strength (Hautmann et al. 2000). It has been shown that swimming training increases in lung volumes because respiratory muscles overcome the increased pressure exerted by water during the breathing cycle. This leads to functional muscle improvement and alterations in lung elasticity, contributing to greater increases in forced vital capacity and other lung function variables in swimmers than in runners (Sable et al. 2012). Other authors confirm this assumption in a study on disabled swimmers (Okrzymowska et al. 2019). Contrarily, some studies support the notion that competitive swimming does not affect lung growth during puberty, and larger lungs in swimmers are an inherent feature rather than induced by swimming (Bovard et al. 2018). It is not full well settled to what extend a greater lung volume in swimmers is a consequence of training (Clanton et al. 1987), genetic endowment (Bar-Or et al. 1994), or increased respiratory muscle strength (Doherty and Dimitriou 1997). In this context, it is worth noting that 14 swimmers out of the 28 endurance athletes in the present study demonstrated the highest values of MIP and MEP and of spirometry variables.

We further found in this study positive correlations between respiratory muscle strength, expressed by MIP and MEP, and ventilatory variables. These correlations were of variable strength, being more expressive for MEP than MIP, and the strongest was for both MIP and MEP in relation to MVV. However, the correlations were present in power athletes and control untrained subjects, but not in endurance athletes. MVV is considered an indicator of forced respiratory power or functional reserve of respiratory muscles as well as of biomechanical capability of lungs and airways. The corollary is that ventilatory function goes together with respiratory muscle strength in power athletes and untrained subjects, but not in well-trained endurance athletes, who, incidentally, showed the greatest increases in MIP, MEP, and in spirometry variables. Thus, it is probable that intensive endurance training, particularly swimming, results in achieving the upper limits of functional reserves of respiratory muscles and ventilatory function. This notion finds support also in other studies. Klusiewicz (2008) has shown that well-trained endurance athletes at a point of training reach the maximum inspiratory muscle strength, and further training would not further improve muscle strength. It has also been shown that flow-resistive inspiratory muscle training is not more effective for respiratory muscle improvement than a competitive swimming training program in elite swimmers (Mickleborough et al. 2008). However, inspiratory muscle training in conjunction with swimming training improves respiratory muscle function in sub-elite swimmers when compared to swimming alone (Shei et al. 2016), or in young 12-year-old fin-swimmers (Vašíčková et al. 2017). Possibly, the best effect of a combination of regular training and inspiratory muscle training may be achieved in young athletes during the time of active growth, like 10–16 years of age. That suggestion requires further exploratory research taking into consideration age at training onset, intensity of exercise, and a specific genetic endowment. It would be more informative to investigate the relationship between respiratory muscle strength and ventilatory function in athletes involved in different endurance and power sports, e.g., swimming and power lifting. The results of the present study are not cogent to this end due to a small number of well-trained athletes participating. Nonetheless, the present findings show that type of sporting activity bears on the physiological adaptation of respiratory system, inclusive of respiratory muscle strengths. Stronger function of respiratory muscles provides greater lung ventilation during exercise and greater maximum voluntary ventilation. When the inspiratory muscles are well trained, triggering of the metaboreflex, which limits blood flow to the working muscles and accelerates muscle fatigue, is delayed and performance increases.

We conclude that regular endurance training leads to the most adaptive and prominent changes in respiratory muscle strength and lung function compared to specific power training; the latter still yielding a better lung function than that present in untrained persons. Additionally, we believe we have shown that intensive endurance training, exemplified by swimming, may lead to the ceiling for improvement of respiratory muscle strength, and correspondingly in ventilatory function, with no reserve left over. On the other hand, targeted muscle training may be an effective strategy to improve respiratory muscle function and ventilation in power athletes, especially in athletes with low inspiratory/expiratory pressure, as well as in less physically fit healthy persons.

Acknowledgments Our thanks go to the Military Sport Institute in St. Petersburg, Russia, for the possibility to conduct research and to the athletes who agreed to participate in the study.

Conflicts of Interest The authors declare no conflict of interest in relation to this article.

Ethical Approval All procedures performed in studies involving human participants were in accordance with the ethical standards of the institutional and/or national research committee and with the 1964 Helsinki declaration and its later amendments or comparable ethical standards.

Experimental procedures were approved by the Ethics Committee of the Pavlov Institute of Physiology of the Russian Academy of Sciences in St. Petersburg, Russia, conducted in accordance with the World Medical Association Declaration of Helsinki.

Informed Consent Written informed consent was obtained from all the individual participants included the study.

References

Armour J, Donnelly PM, Bye PT (1993) The large lungs of elite swimmers: an increased alveolar number? Eur Respir J 6:237–247

ATS/ERS (2002) American Thoracic Society/European Respiratory Society. Statement on respiratory muscle testing. Am J Respir Crit Care Med 166:518–624

Bar-Or O, Unithan V, Illescas C (1994) Physiologic considerations in age-group swimming. Med Sci Aquatic Sports 39:199–205

Boutellier U, Büchel R, Kundert A, Spengler C (1992) The respiratory system as an exercise limiting factor in normal trained subjects. Eur J Appl Physiol Occup Physiol 65:347–353

Bovard JM, Welch JF, Houghton KM, McKenzie DC, Potts JE, Sheel AW (2018) Does competitive swimming affect lung growth? Physiol Rep 6(15):e13816

Clanton T, Dixon GF, Drake J, Gadek JE (1987) Effects of swim training on lung volumes and inspiratory muscle conditioning. J Appl Physiol 62:39–46

Doherty M, Dimitriou L (1997) Comparison of lung volume in Greek swimmers, land based athletes, and sedentary controls using allometric scaling. Br J Sports Med 31:337–341

HajGhanbari B, Yamabayashi C, Buna TR, Coelho JD, Freedman KD, Morton TA, Palmer SA, Toy MA, Walsh C, Sheel AW, Reid WD (2013) Effects of respiratory muscle training on performance in athletes: a systematic review with meta-analyses. J Strength Cond Res 7:1643–1663

Hautmann H, Hefele S, Schotten K, Huber RM (2000) Maximum inspiratory mouth pressures (PIMAX) in healthy subjects – what is the lower limit of normal? Respir Med 94:689–693

Illi SK, Held U, Frank I, Spengler CM (2012) Effect of respiratory muscle training on exercise performance in healthy individuals. Sport Med 42:707–724

Janssens L, Brumagne S, McConnell AK, Raymaekers J, Goossens N, Gayan-Ramirez G, Hermans G, Troosters T (2013) The assessment of inspiratory muscle fatigue in healthy individuals: a systematic review. Respir Med 107:331–346

Johnson BD, Aaron EA, Babcock MA, Dempsey JA (1996) Respiratory muscle fatigue during exercise: implications for performance. Med Sci Sport Exerc 28:1129–1137

Klusiewicz K (2008) Characteristics of the inspiratory muscle strength in the well-trained male and female athletes. Biol Sport 25:13–22

Klusiewicz K, Borkowski L, Zdanowicz R, Boros P, Wesolowski S (2008) The inspiratory muscle training in elite rowers. J Sport Med Phys Fit 48:279–284

Lazovic B, Mazic S, Suzic-Lazic J, Djelic M, Djordjevic-Saranovic S, Durmic T, Zikic D, Zugic V (2015) Respiratory adaptations in different types of sport. Eur Rev Med Pharmacol Sci 19:2269–2274

Lazovic-Popovic B, Zlatkovic-Svenda M, Durmic T, Djelic M, Djordjevic Saranovic S, Zugic V (2016) Superior lung capacity in swimmers: some questions, more answers! Rev Port Pneumol 22:151–156

Mickleborough TD, Stager JM, Chatham K, Lindley MR, Ionescu AA (2008) Pulmonary adaptations to swim and inspiratory muscle training. Eur J Appl Physiol 103:635–646

Miller MR, Crapo R, Hankinson J, Brusasco V, Burgos F, Casaburi R, Coates A, Enright P, van der Grinten CP, Gustafsson P, Jensen R, Johnson DC, MacIntyre N, McKay R, Navajas D, Pedersen OF, Pellegrino R, Viegi G, Wanger J, ATS/ERS Task Force (2005) General considerations for lung function testing. Eur Respir J 26:153–161

Neves LF, Reis MH, Plentz RD, Matte DL, Coronel CC, Sbruzzi G (2014) Expiratory and expiratory plus inspiratory muscle training improves respiratory muscle strength in subjects with COPD: systematic review. Respir Care 59:1381–1388

Okrzymowska P, Kurzaj M, Seidel W, Rozek-Piechura K (2019) Eight weeks of inspiratory muscle training improves pulmonary function in disabled swimmers – a randomized trial. Int J Inv Res Public Health 16:1747

Ozmen T, Gunes GY, Ucar I, Dogan H, Gafuroglu TU (2017) Effect of respiratory muscle training on pulmonary function and aerobic endurance in soccer players. J Sport Med Phys Fitness 57:507–513

Romer L, Dempsey J (2006) Legs play out for the cost of breathing! Physiol News 65:25–27

Romer LM, Polkey MI (2008) Exercise-induced respiratory muscle fatigue: implications for performance. J Appl Physiol 104:879–888

Rong C, Bei H, Yun M, Yuzhu W, Mingwu Z (2008) Lung function and cytokine levels in professional athletes. J Asthma 45:343–348

Rosser-Stanford B, Backx K, Lord R, Williams EM (2019) Static and dynamic lung volumes in swimmers and their ventilatory response to maximum exercise. Lung 197:15–19

Sable M, Vaidya SM, Sable SS (2012) Comparative study of lung functions in swimmers and runners. Indian J Physiol Pharmacol 56:100–104

Sales AT, Fregonezi GA, Ramsook AH, Guenette JA, Lima IN, Reid WD (2016) Respiratory muscle endurance after training in athletes and non-athletes: a systematic review and meta-analysis. Phys Ther Sport 17:76–86

Segizbaeva MO, Donina ZA, Timofeev NN, Korolyov YN, Golubev VN, Aleksandrova NP (2013) EMG analysis of human inspiratory muscle resistance to fatigue during exercise. Adv Exp Med Biol 788:197–205

Segizbaeva MO, Timofeev NN, Donina ZA, Kur'yanovich EN, Aleksandrova NP (2015) Effects of inspiratory muscle training on resistance to fatigue of respiratory muscles during exhaustive exercise. Adv Exp Biol Med 840:35–43

Shei RJ, Lindley M, Chatham K, Mickleborough TD (2016) Effect of flow–resistive inspiratory loading on pulmonary and respiratory muscle function in sub–elite swimmers. J Sports Med Phys Fitness 56:392–398

Tiller NB, Simpson AJ (2018) Effect of spirometry on intra-thoracic pressures. BMC Res Notes 11(1):110

Troosters T, Gosselink R, Decramer M (2005) Respiratory muscle assessment. In: Gosselink R, Stam H (eds) Lung function testing. European respiratory monograph, vol 31. European Respiratory Society Journals Ltd, Wakefield/Sheffield, pp 57–71

Vašíčková J, Neumannová K, Svozil Z (2017) The effect of respiratory muscle training on fin-swimmers' performance. J Sports Sci Med 16:521–526

Wells GD, Norris SR (2009) Assessment of physiological capacities of elite athletes and respiratory limitations to exercise performance. Pediatr Respir Rev 10:91–98

Witt JD, Guenette JA, Rupert JL, McKenzie DC, Sheel AW (2007) Inspiratory muscle training attenuates the human respiratory muscle metaboreflex. J Physiol 584:1019–1028

Wüthrich TU, Notter DA, Spengler CM (2013) Effect of inspiratory muscle fatigue on exercise performance taking into account the fatigue-induced excess respiratory drive. Exp Physiol 98:1705–1717

Adv Exp Med Biol - Clinical and Experimental Biomedicine (2021) 11: 99–105
https://doi.org/10.1007/5584_2020_543

Published online: 10 July 2020

Uroflowmetry and Altitude Hypoxia: A Report from Healthy Italian Trekkers and Nepali Porters During Himalayan Expedition

Vittore Verratti, Danilo Bondi, Aliasger Shakir, Tiziana Pietrangelo, Raffaela Piccinelli, Vincenzo Maria Altieri, Danilo Migliorelli, and Alessandro Tafuri

Abstract

Hypoxia alters micturition, which influences bladder function by involving different neurological and humoral systems. In this study we assessed the mid-term effects of altitude hypoxia on uroflowmetry in healthy male lowlander native Nepali porters and Italian trekkers, four each, who coattended a Himalayan expedition. All the participants completed a 19-day trek along a demanding route with ascent and descent at the Kanchenjunga Mountain. They underwent micturition and urodynamic analysis twice, at low altitude of 665 m a.s.l. and high altitude of 4,750 m a.s.l. Statistical comparisons considered the altitude effects (low vs. high) and ethnicity (Italian vs. Nepali). Food consumption was recorded, and water and energy intake were calculated. We found trends of borderline significance in the mean urinary flow rate (Q_{mean}) ($p = 0.058$; effect size $\eta^2 p = 0.478$) and in Q_{max} to the advantage of the Nepali. There was no evidence of differences when comparing time to Q_{max} and urine volume at Q_{max} and Q_{mean} for altitude or altitude × ethnicity. In addition, there was a lonely female participant, who, analyzed as a case report, showed increased Q_{mean} at high altitude. Older age mitigated while energy

V. Verratti (✉)
Department of Psychological, Health and Territorial Sciences, University "G. d'Annunzio" of Chieti-Pescara, Chieti, Italy
e-mail: vittore.verratti@unich.it

D. Bondi and T. Pietrangelo
Department of Neuroscience, Imaging and Clinical Sciences, University "G. d'Annunzio" of Chieti-Pescara, Chieti, Italy

A. Shakir
USC Institute of Urology and Catherine and Joseph Aresty Department of Urology, Keck School of Medicine, University of Southern California (USC), Los Angeles, CA, USA

R. Piccinelli
Research Center for Food and Nutrition, Council for Agricultural Research and Economics, Rome, Italy

V. M. Altieri
Department of Urology, Humanitas Gavazzeni Hospital, Bergamo, Italy

D. Migliorelli
COMPUMED Europe, Rome, Italy

A. Tafuri
Department of Neuroscience, Imaging and Clinical Sciences, University "G. d'Annunzio" of Chieti-Pescara, Chieti, Italy

USC Institute of Urology and Catherine and Joseph Aresty Department of Urology, Keck School of Medicine, University of Southern California (USC), Los Angeles, CA, USA

Department of Urology, University of Verona, Azienda Ospedaliera Universitaria Integrata Verona, Verona, Italy intake potentiated the ethnic differences noted in uroflowmetry. We conclude that altitude hypoxia rather inappreciably affects micturition in healthy men. However, a trend for possible ethnic differences raises worthy of note perspectives on adaptive ability of micturition. Also, dietary intake and age should be considered as confounding elements when evaluating micturition.

Keywords

Altitude hypoxia · Bladder · Himalayan expedition · Hypobaric hypoxia · Micturition · Uroflowmetry

1 Introduction

High-altitude hypoxia model is usually adopted to investigate the effects of low oxygen tension on physiological functions (Verratti et al. 2016a, b). Hypoxia also is a frequent age-related condition, and hypoxic models can be used to simulate organ aging (Cataldi and Di Giulio 2009). It has been shown that aging alters micturition, notably causing detrusor muscle dysfunction related to oxidative stress (Van Haarst et al. 2004). Hypoxia alters micturition by direct and indirect actions involving different neurological and humoral systems (Verratti et al. 2019; Scheepe et al. 2011).

Hypoxia is present in the bladder microenvironment when intravesical pressure increases due to bladder outflow obstruction, a frequent occurrence in prostatic enlargement (Verratti et al. 2016a, b; Mehrotra 1953). This is an age-related condition that worsens quality of life (Kupelian et al. 2006) and which is ameliorated by relieving the outflow obstruction that increases oxygen supply to the detrusor muscle (Morelli et al. 2011). However, the influence of hypoxia on uroflowmetry dynamics is not full clear. We have recently demonstrated in young adult women that micturition is affected during adaptation to hypoxia (Verratti et al. 2019). In contradistinction, Ohno et al. (2018) have failed to find pathological urodynamic changes in the long-term hypoxic exposure during an Antarctic expedition.

A recent growing interest in high-altitude tourism spurs a need for the investigation of specific urological risks of altitude sojourn. In this context, urinary function comparison in lowlander trekkers and local native porters during an expedition represents a hypoxic scenario of adaptation that takes place at high altitude depending on ethnicity. Therefore, the aim of this study was to assess the mid-term effects of altitude hypoxia on male uroflowmetry, comparing Italian trekkers and native Nepali porters who coattended a Himalayan expedition.

2 Methods

This study was part of a research project entitled "Kanchenjunga Exploration and Physiology", belonging to a broader scientific undertaking termed "Environmentally-modulated metabolic adaption to hypoxia in altitude natives and sea-level dwellers: from integrative to molecular level". There were four Italian trekkers (mean age 41 $\pm$ 15 years, body mass index (BMI) 25.2 $\pm$ 3.7 kg/m^2) and four Nepali porters (mean age 29 $\pm$ 8 years, BMI 26.3 $\pm$ 4.1 kg/m^2) included into the study. In addition to these men, there was one female trekker participating in the expedition (age 36 years, BMI 25.1 kg/m^2). All the participants were free of any urological ailments. Only the men were included in the detailed analysis. The inclusion criteria were non-forced micturition and a total urine volume, decreased by (minus residual volume), greater than 150 mL. The female participant was in here presented separately as a case report. Participants completed a combined circuit of 300 km distance in 19 days (Fig. 1), a summary difference in altitude of over 16,000 m, covering a daily average of 6 h walk, along a demanding route with ascent and descent in the Himalayas of Nepal.

Food consumption was self-recorded by the participants, after being instructed by an expert field worker, for 2 or 3 non-consecutive days. Food and beverages ingested were recorded in two hard copy diaries structured by seven meals (three main meals and four snacks). Quantities were specified referring to the standard household measures or standard portions of a dedicated picture atlas. The field worker checked and entered all the diary information into web-based software "Food Consumption Database" (FOODCONS).

Four sets of databases (food descriptors, household units of measurement, standard recipes, and food composition) were used to transform the food data into the mass content of single raw ingredients and into the amount of nutrients consumed. FOODCONS software and all related instruments, such as food diaries and picture atlas, were adopted from those developed by the Research Center for Food and Nutrition of the Council for Agricultural Research and Economics (CREA) in Roma for use in the Fourth National Dietary Survey in Italy (SCAI 2017). In the present study, only the average daily intakes of energy and water were reported. On average, daily water intake of the Italians was 3.14 ± 0.52 L, of which 67% came from beverages and 33% from foods, while that of the Nepalis was 3.30 ± 0.41 L, of which 64% came from beverages and 36% from foods. Daily energy intake was 2,898 ± 897 kcal for the Italians and 2,816 ± 226 kcal/day for the Nepalis.

Additionally, we measured peripheral blood oxygen saturation (SpO_2) in both groups of trekkers at low and high altitudes (APN-100 Monitor; Intermed, Milano, Italy). These measurements were done immediately before the assessment of uroflowmetry that was performed twice: at low (Dobhan, 665 m a.s.l.) and high altitude (Lhonak, 4750 m a.s.l.) (Fig. 1). Pre-micturition volume and post-voiding residual volume were evaluated with Padscan HD5 Bladder Scanner (Caresono Technology; Nanshan, Shenzhen, Guangdong, China). Uroflowmetry variables were obtained using a SmartScale wireless urine flowmeter (Albyn Medical; Cordovilla, Navarra, Spain). In detail, we assessed the following urinary flow rates: Q_{max}, Q_{mean}, time to Q_{max}, and volume at Q_{max}. The very nature of the present field study in an extreme outdoor environment could have biased other uroflowmetry variables, e.g., micturition volume and time, or urine flowtime, which therefore were not considered. Participants were asked to report their urinary habits during the days of ascent.

Results were expressed as means ±SD. Data distribution was checked with the Shapiro-Wilk test. Levene's test was used for the assessment of variance homogeneity. Further, consideration was given to the issues of sphericity violation, unbalanced design, and missing data (Armstrong 2017). Repeated measures ANOVA was performed to compare differences between low vs. high altitude and between low vs. high altitude × ethnic group (Italian vs. Nepali). Then, the analysis was repeated after inclusion age, BMI, and water and energy intake as covariates.

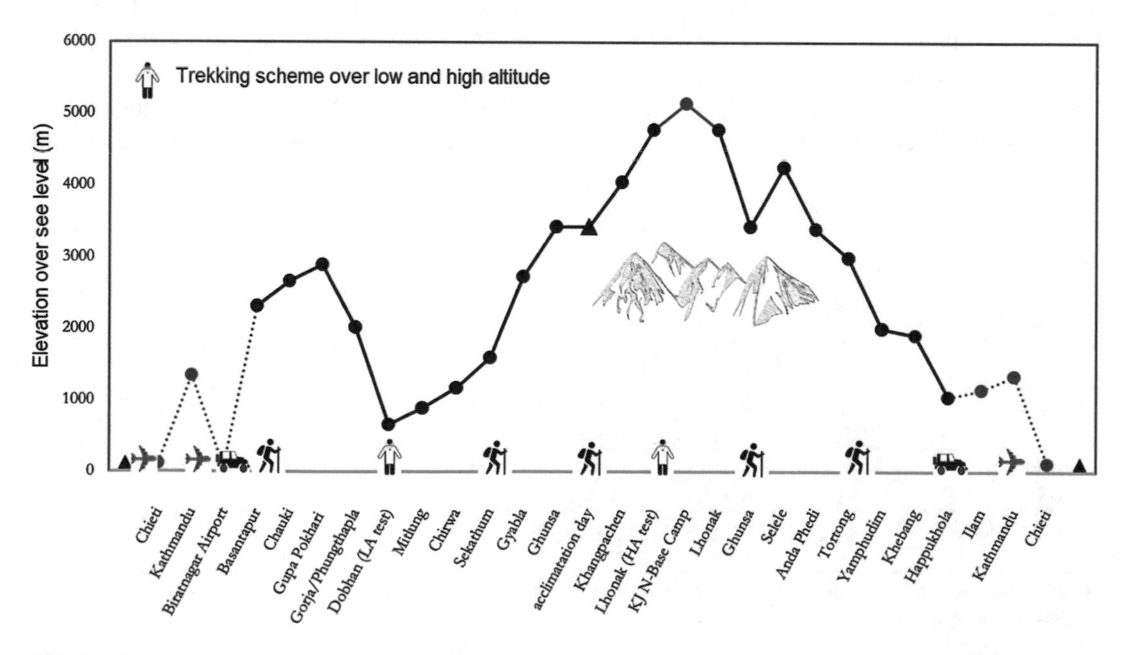

Fig. 1 Altimetric scheme of Himalayan trekking expedition, part of the "Kanchenjunga Exploration and Physiology project"

An alpha level of <0.05 was used to define statistical significance and effect size (partial η^2).

3 Results

We failed to substantiate the presence of any significant differences in urine flow variables between the Italian and native Nepali porters, at low and high altitude, or the interaction of altitude × ethnicity. The only insignificant differences we noticed were tendencies for a larger urine Q_{max} and volume at Q_{max} in the Nepalis ($p = 0.106$; $\eta^2 = 0.376$ and $p = 0.119$; $\eta^2 = 0.355$, respectively); the effects were present at both low and high altitude (Table 1). However, after correcting for age, the former tendency was mitigated ($p = 0.275$; $\eta^2 = 0.231$). Likewise, there appeared a strong trend for ethnicity in case of Q_{mean}, with the Nepalis having higher values ($p = 0.058$; $\eta^2 = 0.478$) (Fig. 2). Again, after correcting for age, this tendency was mitigated ($p = 0.174$: $\eta^2 = 0.334$), while after correcting for energy intake, it was enhanced to become significant ($p = 0.041$; $\eta^2 = 0.598$). No other significant results emerged after correcting for covariates. Water intake did not affect the results either.

The Italian trekkers reported a higher frequency and shorter duration of micturition with increasing altitude. The results of SpO_2 were as follows: low altitude, Italians 98.0 ± 0.8% and Nepalis 97.3 ± 2.1%; high altitude, Italians 86.5 ± 5.2 and Nepalis 84.0 ± 1.4.

The only one female trekker participating in the expedition had the following results of measurements: Q_{time} = 16.6/12.6 mL/s, time to Q_{max} = 11.3/6.4 s, Q_{max} = 33.5/31.8 mL/s, volume at Q_{max} = 162/123 mL, and Q_{mean} = 12.2/18.1 mL/s at low/high altitude, respectively (Fig. 3).

4 Discussion

There is still a lack of comprehension concerning micturition physiology in response to hypobaric hypoxia. We have previously reported some results in healthy women, with differences in urine flow and volume at high altitude (Verratti et al. 2016a, b). In the present study, we report results from testing healthy males at low and high altitudes while simultaneously investigating the potential role of high-altitude ethnicity in urodynamic changes. We found that altitude failed to appreciably affect uroflowmetry and there was no evidence for any ethnic differences between native Nepali porters and Italian sojourners. Nonetheless, there appeared a general tendency for higher maximum and mean urine flow rates and for urine volume at maximum flow rate in Nepalis when compared to Italians. Further studies with larger samples of trekkers would be needed to get a deeper insight into this observation.

The Nepalis had better bladder compliance (greater volume at maximum flow rate) and

Table 1 Uroflowmetry parameters

Uroflowmetry	Altitude	Ethnicity: Italian	Ethnicity: Nepali	Low vs. high altitude: p	$\eta^2 p$	Italian vs. Nepali: p	$\eta^2 p$	Interaction: p	$\eta^2 p$
Q_{max} (mL/s)	LA	19.8 ± 6.5	29.6 ± 10.9	0.763	0.016	0.106	0.376	0.565	0.058
	HA	20.6 ± 27.2	27.2 ± 5.6						
Q_{mean} (mL/s)	LA	9.6 ± 3.6	16.7 ± 6.6	0.596	0.050	0.058	0.478	0.520	0.072
	HA	11.9 ± 3.7	16.5 ± 2.9						
Time to Q_{max} (s)	LA	8.3 ± 3.1	8.9 ± 2.9	0.219	0.239	0.510	0.075	0.448	0.099
	HA	5.6 ± 1.6	8.1 ± 5.6						
Volume at Q_{max} (mL)	LA	93.1 ± 50.8	151.3 ± 38.2	0.740	0.020	0.119	0.355	0.899	0.003
	HA	88.9 ± 41.2	141.8 ± 69.7						

Q, urine flow rate; *LA*, low altitude; *HA*, high altitude, *p*, significance; *$\eta^2 p$*, partial eta squared: effect size; *interaction*, altitude × ethnicity

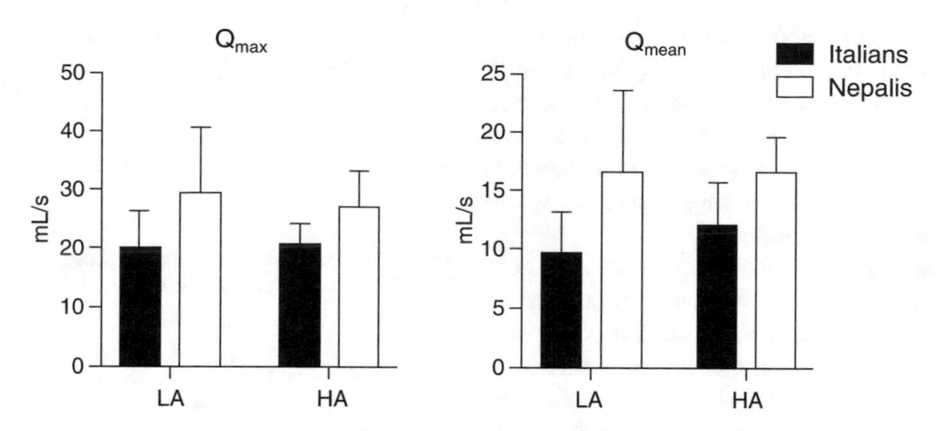

Fig. 2 Uroflowmetry results: maximum (Q_{max}) and mean (Q_{mean}) urine flow at low (LA) and high altitude (HA) in Italian trekkers and native Nepali porters. Data are means +SD

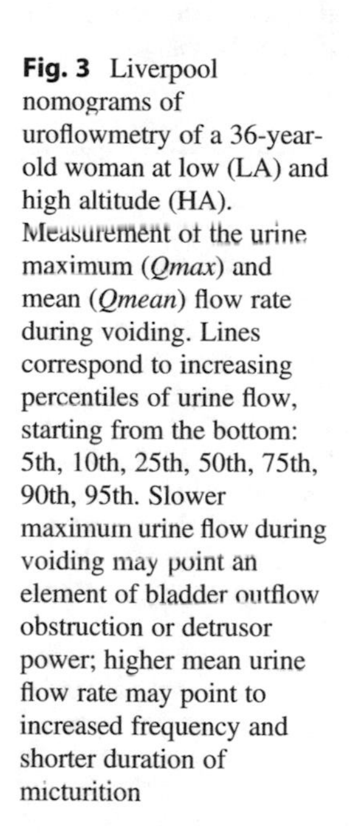

Fig. 3 Liverpool nomograms of uroflowmetry of a 36-year-old woman at low (LA) and high altitude (HA). Measurement of the urine maximum (*Qmax*) and mean (*Qmean*) flow rate during voiding. Lines correspond to increasing percentiles of urine flow, starting from the bottom: 5th, 10th, 25th, 50th, 75th, 90th, 95th. Slower maximum urine flow during voiding may point an element of bladder outflow obstruction or detrusor power; higher mean urine flow rate may point to increased frequency and shorter duration of micturition

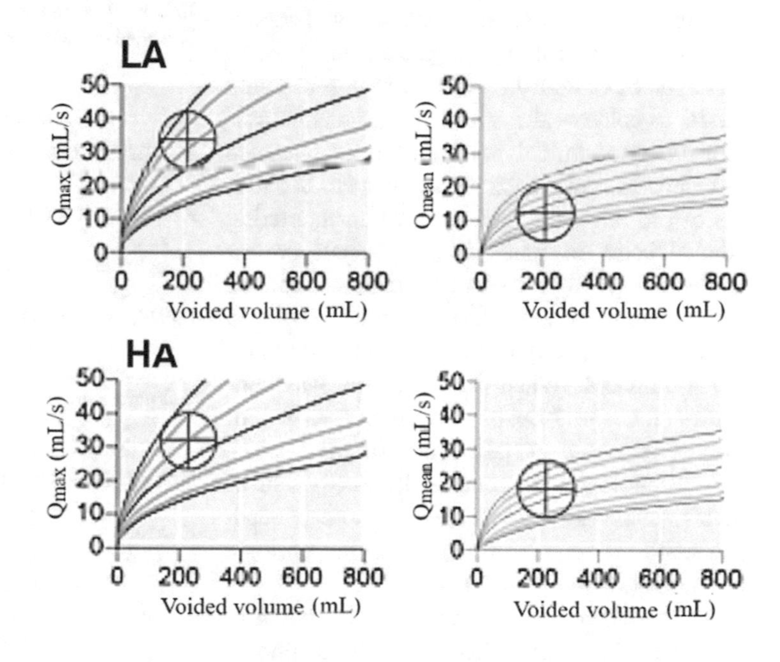

detrusor power (greater maximum and mean flow rate) compared to the Italians, although with no appreciable difference between low and high altitude. These changes likely reflect a better detrusor muscle adaptation to frequent exposures to hypoxia, kind of "hypoxic memory". Indeed, we have previously demonstrated that high altitude provokes a reduction in bladder power at maximum flow (Verratti et al. 2019). The detrusor muscle is under the control of the autonomic nervous system, with sympathetic-driven relaxation and parasympathetic-driven contractility. Thus, it is a reasonable assumption that hypoxia-related downregulation of detrusor activity may be due to a diminished neural input to the muscular bladder compartment, which results in uncoordinated delivery, collection, and expulsion of urine (Hill 2015). Changes in micturition may also be related to other factors, such as diuretic reaction (Goldfarb-Rumyantzev and Alper 2014) or the levels of bladder power at maximum flow and bladder contractility index (Verratti et al. 2019). The trekkers of this study indeed reported an altered urinary habit in higher altitudes, with increased frequency and reduced duration, which might be due both to enhanced altered urge sensation and diuretic response.

Gender and age-related differences have been previously described in voiding function (Verratti et al. 2016a, b; Van Haarst et al. 2004). We were unable to address such differences in this study having just a lonely female sojourner. Nonetheless, we observed in the woman a rather clear increase in the mean flow rate with somehow slower maximum urine flow (Fig. 2), suggesting the possibility of increased frequency and shorter duration of micturition at high altitude. It seems that hypoxia of high altitude may represent a field model to enhance the knowledge of pathophysiology of urine flow dynamics, with the plausible reference to urine incontinence as well. We also noticed in this study that age was an interfering ethnic factor in that younger Nepali porters tended to have higher urine flow rates and volume at maximum flow rate when compared to Italians. Additionally, dietary energy intake affected ethnic differences. When dietary intake was controlled for, the Nepalis had higher values of mean flow rate than Italians did. In this context, new technologies that enable the assessment of dietary intake give perspective to more in-depth research on ethnic differences in urologic functions (Turrini et al. 2019).

In conclusion, we believe we have shown in this study that native Nepali Himalayan porters tend to have enhanced typical uroflowmetry variables at sea level than those present in Caucasian sojourners. Interestingly, altitude hypoxia did not alter this basic ethnic difference in urinary function. Further exploration of ethnic dependency of high-altitude hypoxia on urinary function is needed in larger samples of participants, clustered by gender and controlled for dietary intake to provide deeper insights into the pathophysiology of urinary function.

Acknowledgments Supported by a grant from the Department of Psychological, Health and Territorial Sciences, "G. d'Annunzio" University of Chieti-Pescara, Italy. Our thanks go to the Mission Nepal Holidays Pvt. Ltd. in Kathmandu, Nepal, for the logistic support during the expedition. We also thank the Italian trekkers and Nepali porters involved in this project. A special thank goes to COMPUMED Europe S.R.L. in Rome, Italy, for free instrumental support.

Competing Interests The authors declare no competing interests in relation to this article.

Ethical Approval Approved by the Ethical Review Board of the Nepal Health Research Council (NHRC).

All study procedures were performed in accordance with the ethical standards of the 1964 Helsinki declaration and its later amendments.

Informed Consent All participants provided written informed consent for participation in the study.

References

Armstrong RA (2017) Recommendations for analysis of repeated-measures designs: testing and correcting for sphericity and use of manova and mixed model analysis. Ophthalmic Physiol Opt 37(5):585–593

Cataldi A, Di Giulio C (2009) Oxygen supply as modulator of aging processes: hypoxia and hyperoxia models for aging studies. Curr Aging Sci 2(2):95–102

Goldfarb-Rumyantzev AS, Alper SL (2014) Short-term responses of the kidney to high altitude in mountain climbers. Nephrol Dial Transplant 29(3):497–506

Hill WG (2015) Control of urinary drainage and voiding. Clin J Am Soc Nephrol 10(3):480–492

Kupelian V, Wei JT, O'Leary MP, Kusek JW, Litman HJ, Link CL, McKinlay JB, Survery Investigators BACH (2006) Prevalence of lower urinary tract symptoms and effect on quality of life in a racially and ethnically diverse random sample: the Boston Area Community Health (BACH) survey. Arch Intern Med 166 (21):2381–2387

Mehrotra R (1953) An experimental study of the vesical circulation during distension and in cystitis. J Pathol Bacteriol 66(1):79–89

Morelli A, Sarchielli E, Comeglio P, Filippi S, Mancina R, Gacci M, Vignozzi L, Carini M, Vannelli GB, Maggi M (2011) Phosphodiesterase type 5 expression in human and rat lower urinary tract tissues and the effect of tadalafil on prostate gland oxygenation in spontaneously hypertensive rats. J Sex Med 8(10):2746–2760

Ohno G, Otani S, Ikeda A (2018) Human beings in Antarctica – a key to global change. Book chapter; IntechOpen – 10 Lower Thames Street, London, EC3R 6AF, UK. https://doi.org/10.5772/intechopen.81974.; https://www.intechopen.com/books

SCAI (2017) IVth National Dietary Survey in Italy. https://healthydietforhealthylife.eu/index.php/news/355-launch-of-the-iv-national-dietary-survey-in-italy-iv-scai. Accessed on 14 Apr 2020

Scheepe JR, Amelink A, de Jong BD, Wolffenbuttel KP, Kok DJ (2011) Changes in bladder wall blood oxygen saturation in the overactive obstructed bladder. J Urol 186(3):1128–1133

Turrini A, D'Addezio L, Dhurandhar EJ, Ferrari M, Le Donne C, Mistura L, Piccinelli R, Scalvedi ML, Sette S (2019) Emerging topics in dietary assessment. Front Nutr 6:176

Van Haarst EP, Heldeweg EA, Newling DW, Schlatmann TJ (2004) The 24-h frequency-volume chart in adults reporting no voiding complaints: defining reference values and analysing variables. BJU Int 93 (9):1257–1261

Verratti V, Di Giulio C, D'angeli A, Tafuri A, Francavilla S, Pelliccione F (2016a) Sperm forward motility is negatively affected by short-term exposure to altitude hypoxia. Andrologia 48(7):800–806

Verratti V, Paulesu L, Pietrangelo T, Doria V, Di Giulio C, Aloisi AM (2016b) The influence of altitude hypoxia on uroflowmetry parameters in women. Am J Physiol Renal Physiol 311(3):F562–F566

Verratti V, Mrakic-Sposta S, Moriggi M, Tonacci A, Bhandari S, Migliorelli D, Bajracharya A, Bondi D, Agrò EF, Cerretelli P (2019) Urinary physiology and hypoxia: a pilot study of moderate-altitude trekking effects on urodynamic indexes. Am J Physiol Renal Physiol 317(4):F1081–F1086

Adv Exp Med Biol - Clinical and Experimental Biomedicine (2021) 11: 107–114
https://doi.org/10.1007/5584_2020_551

Published online: 14 July 2020

Effects of Intraligamentary Injection of Osteogenic-Induced Gingival Fibroblasts on Cementum Thickness in the Dog Model of Tooth Root Resorption

Tarek El-Bialy, Hagai Hazan Molina, Yuval Aizenbud, Wasif Qayyum, Saleem Ali, and Dror Aizenbud

Abstract

Tooth root resorption is an unwanted result of orthodontic tooth movement, and it can be expressed by a reduction in cementum thickness. The aim of this experimental study was to evaluate the effect of intraligamentary injection of osteogenic-induced gingival fibroblasts (OIGF) on cellular and acellular tooth root cementum thickness in modeled orthodontic tooth movement. Six beagle dogs were used in the study. All the upper and lower third and fourth premolars were subjected to mechanical loading for 4 weeks, which induced orthodontic tooth movement. Fifteen premolars were assigned to the OIGF group, which received a single OIGF injection through the periodontal ligament near the root apex ($n = 7$ teeth), and to the control group, which received a single injection of Dulbecco's modified eagle's medium in the periapical area ($n = 8$ teeth). The evaluation of histomorphometry was performed to assess the thicknesses of cellular and acellular cementum at the root apex and four bilateral sites distal to the apex. We found no statistically significant enhancing effects of gingival fibroblasts on either cellular or acellular cementum thicknesses when compared with the control group. We conclude that a single intraligamentary injection of OIGF does not stimulate the formation of tooth root cementum in the dog model of orthodontic tooth movement. Thus, OIGF is unlikely to prevent orthodontic-induced tooth root resorption.

T. El-Bialy
Division of Orthodontics, School of Dentistry, Faculty of Medicine and Dentistry, University of Alberta, Edmonton, AB, Canada

H. Hazan Molina and D. Aizenbud (✉)
Department of Orthodontics and Craniofacial Anomalies, School of Graduate Dentistry, Rambam Health Care Center and Technion Haifa, Haifa, Israel
e-mail: aizenbud@ortho.co.il

Y. Aizenbud
Institute of Dental Sciences, Faculty of Dental Medicine, Hebrew University, Jerusalem, Israel

W. Qayyum
Division of Oral Biology, School of Dentistry, Faculty of Medicine and Dentistry, University of Alberta, Edmonton, AB, Canada

S. Ali
Department of Orthodontics, Queen Medical Center, Doha, Qatar

Keywords

Gingival fibroblasts · Orthodontic tooth movement · Premolars · Tooth cementum · Tooth resorption · Tooth root

1 Introduction

Cementum is an avascular, mineralized connective tissue layer that covers the tooth root (Bosshardt and Selvig 1997). It is different from bone as it increases in thickness throughout life and does not undergo dynamic remodeling. Historically, cementum is classified as acellular cementum, a thin coverage of the cervical root which does not include cementocytes, and a thick cellular cementum covering the apical root, which includes cementocytes (Yamamoto et al. 2016). The cementum is responsible for anchoring the tooth root surface to the alveolar bone through periodontal ligament fibers (Grzesik and Narayanan 2002). Cementum has two types of fibers, extrinsic Sharpey's fibers, which are the embedded ends of principal fibers secreted by fibroblasts and partly by cementoblasts, and intrinsic fibers secreted by cementoblasts (Yamamoto et al. 2016). There are mainly two types of cementum: cellular intrinsic fiber cementum or secondary cementum and acellular extrinsic fiber cementum or primary cementum (Foster 2012). Cementum forming cells, the cementoblasts, are continuously recruited from cementoprogenitor cells confined within the resorption lacunae, which are pits formed on the root surface during tooth root resorption (Zeichner-David 2006; Bosshardt and Selvig 1997). An important function of cementum is its repair and adaptive role in the maintenance of the tooth root and hence in the preservation of its supporting apparatus.

Orthodontic-induced tooth root resorption (OITRR) is an unwanted but unavoidable result of tooth movement (Brezniak and Wasserstein 2002). Vasculature compression in the periodontium due to mechanical load leads to ischemia (Chan and Darendeliler 2006). Subsequently, macrophage-like cells, multinucleated cells, osteoclasts, and cementoclasts/odontoclasts actively migrate to the hyalinized periodontal tissue in response to the induction of chemotactic agents. In the root-bone interface, necrotic hyalinized tissue is removed while the additive damaging effect is root surface resorption (Jager et al. 2008). Thereafter, fibroblast-like cementoblastic cells invade from the periodontal ligament into the resorption lacunae giving rise to the early cementum repair process. These cells secrete matrix proteins and subsequently collagen fibrils, which fill the lacuna spaces and integrate with the residual collagen fibers forming a thin cementoid repair structure.

It has been reported that 40% of adults during orthodontic treatment have at least one tooth that demonstrates 2.5 mm or greater OITRR (Mirabella and Artun 1995). Others have reported OITRR in about 80% of subjects undergoing orthodontic treatment (Motokawa et al. 2012). It has been stated that orthodontic force as little as 50 g can result in OITRR in a period of 35 days (Harry and Sims 1982). The severity of OITRR is directly related to compression stress (Chan and Darendeliler 2006), and it is classified into three degrees. The first degree is resorption of the superficial layer of cementum, termed cemental or surface resorption. The second is resorption of the root layer or dentin, termed dentinal or deep resorption. The third is circumferential apical resorption, which is complete resorption of the hard tissue components of the root apex, leading to its shortening, and which adversely affects the crown-to-root ratio (Brezniak and Wasserstein 2002). However, in the vast majority of OITRR cases, a slight reduction in root thickness is clinically insignificant and does not affect the prognosis of the involved teeth (Killiany 1999). Furthermore, on cessation of the orthodontic tooth movement, some repair with cellular cementum occurs (Remington et al. 1989).

Gingival fibroblasts are located at the supracrestal surface, i.e., the most occlusal area of the tooth, just below the gum epithelium. Several studies have shown that gingival fibroblasts produce high levels of osteoprotegerin, a soluble protein which considerably decreases the formation of osteoclasts and thus prevents tooth root resorption (de Vries et al. 2006). Despite the fact that stem cells have been successfully used for repair of periodontal defects (Tobita et al. 2008), a major drawback associated with their use is donor site morbidity. Gingival fibroblasts differentiate into cementoblast-like cells (Dogan et al.

2002) and are more easily accessible compared to stem cells, increasing vascularization in vivo and improving gingival attachment in humans (Mohammadi et al. 2007). In addition, the fibroblasts can differentiate into osteogenic cells to repair lost hard tissues (Mostafa et al. 2008).

The aim of this study was a histomorphometry analysis of the effects of intraligamentary injection of osteogenic-induced gingival fibroblasts (OIGF) on cellular and acellular tooth root cementum thickness in beagle dogs with modeled orthodontic tooth root resorption. Currently, there is no information in the literature regarding the use of intraligamentary injection of OIGF and its possible enhancing effect on the cementum thickness in vivo. We hypothesized that intraligamentary injection of OIGFs could induce the formation of tooth root cellular and acellular cementum.

2 Methods

Six beagle dogs, aged 19 months ±8 days, were obtained from the Marshall Bio Resources of North America through the University of Alberta, Canada. The experiments were performed on premolar teeth. Fifteen premolars (upper and lower) were randomly assigned to the OIGF group, which received a 0.5 mL single autologous OIGF injection near the root apex through the periodontal ligament ($n = 7$ teeth) after 4 weeks of orthodontic tooth movement, and to the control group, which received a single injection of Dulbecco's modified eagle's medium in the periapical area ($n = 8$ teeth). All the premolars underwent orthodontic tooth movement for 4 weeks.

2.1 Preparation of Dogs and the Model of Orthodontic Tooth Movement

The dogs were first premedicated with a sedative and antiemetic (Acepromazie 0.05 mg/kg, SubQ), an analgesic (Hydromorphone 0.1 mg/kg, SubQ), and a muscarinic anticholinergic drug (Glycopyrrolate 0.01 mg/kg, SubQ) and then were intubated for inhalation anesthesia with isoflurane in 100% oxygen for a full crown preparation on the third and fourth premolars. Crown cuttings for the orthodontic tooth movement were completed with a tapered diamond bur, and impressions were taken with polyvinyl siloxane. Inhalation anesthesia was then discontinued, and the dogs were transferred to kennels after restoration of normal breathing. In the dental laboratory during wax-up, GAC Dentsply 0.022″ × 0.28″ bondable molar tubes were glued on the buccal surfaces of all crowns of the third and fourth premolars with straight 0.021″ × 0.025″ stainless steel wires in the tube bracket slots to hold the tubes in the same vertical and horizontal orientation. The crowns were then casted into low-fusing metal alloy, finished, polished, and their fitting surfaces were sandblasted with aluminum oxide particles.

Once the crowns were received from the dental laboratory, the dogs were premedicated, intubated, and anesthetized, as above outlined for crown insertion on the premolars. The crowns were cemented with NX3 Nexus® Third Generation dual cure permanent resin cement system (Brea, CA). The OptiBond all-in-one adhesive system was applied twice (scrubbed with a brushing motion for 20 s) on enamel/dentin of the third and fourth premolars, dried for 5 s with medium air, and then the premolars were light-cured for 10 s. A piece of straight 0.021″ × 0.025″ stainless steel wire was inserted into the attachment tube with an open coil spring that was compressed between the two tubes (on the third and fourth premolars) to deliver a force of 1 N per appliance, which was measured using a force gauge. Orthodontic tooth movements were continued for 4 weeks; the coil springs were evaluated and readjusted each week to maintain a force level of 1 N.

2.2 Osteogenic-Induced Gingival Fibroblasts (OIGF)

Concurrently with the orthodontic tooth movement, when impressions were taken for constructing the crowns, the interdental papilla from the third and fourth premolars was excised and immersed in Dulbecco's Modified Eagle's medium (DMEM), supplemented with 100 μg/mL streptomycin and 100 IU/mL penicillin. Using a sterile scalpel, the papillae were cut into smaller pieces, dispersed on slides, and placed in culture plates containing DMEM supplemented with 2 mM L-glutamine solution, 10% fetal bovine serum (FBS), 100 μg/mL streptomycin, 100 IU/mL penicillin, and a 5% CO_2 overlay at 37 °C until 80% confluence. The cells were detached with a 0.05% trypsin and 0.02% ethylenediaminetetraacetic acid (EDTA) solution for 2.5 min and were subcultured to 48-well plates at a density of 2.5×10^3 cells/well. Then, cells were treated with the osteogenic medium consisting of DMEM, 10% FBS, 2 mM L-glutamine, 100 μg/mL streptomycin, 100 IU/mL penicillin, 100 nM dexamethasone, 10 mM β-glycerophosphate, and 50 μg/mL ascorbic acid. The cells received low-intensity pulsed ultrasound (LIPUS) treatment for 20 min/day for 4 weeks, using an incident intensity of 30 mW/cm^2 of the transducer surface (2.5 cm transducer). In effect, 258 gingival fibroblasts differentiated into osteogenic-259 induced gingival fibroblasts (OIGF). All reagents were purchased from Sigma-Aldrich, MO.

2.3 Tissue Preparation for Analysis

After undergoing treatment, a 22-gauge catheter was placed into the right cephalic vein of each dog. The animals were euthanized by intramuscular injection of Domitor® (medetomidine hydrochloride; 0.25 mg/kg) followed by 0.44–0.67 mL/kg of euthanyl. Clinical death was confirmed by evaluating the vital signs. The mandibles and maxillae were dissected and sectioned, using a bone saw. Each section block contained the third and fourth premolars with the supporting alveolar bone. The samples were stored into freshly prepared 4% paraformaldehyde in labeled containers.

2.4 Histomorphometry and Data Evaluation

Seven teeth were investigated in the OIGF group and 8 teeth in the control group. There were 13 slides evaluated in each group, consisting of 6 mesial and 7 distal OIGF roots and 7 mesial and 6 distal control roots. Images were created via a light microscope and digital camera at x2.52 magnification. The thickness of both cellular and acellular cementum was measured at the following 5 points: root apex, 500 μm and 1000 μm distance from the apex on the right (mesial) side and 500 μm and 1000 μm distance from the apex on the left (distal) side in both OIGF and control groups; a total of 10 points of measurement per group. The measurement results were shown in Figs. 1 and 2, respectively. The thickness was first measured by drawing a line from top to bottom of the layer and then by measuring that line. The measurement results were put down in excel spreadsheet for further analysis. The group assessment was blinded. Histomorphometry analysis was performed using Olympus® cellSens software for life science imaging (Olympus; Tokyo, Japan).

Data on the cementum thickness at various measurements points were organized according to a labeling key and were expressed as means ±SE. Quantitative differences between the corresponding points in the active treatment (OIGF) and control groups were evaluated using a two-sample *t*-test. Levene's test was used to assess the equality of variances for the corresponding variables in the two groups. A p-value <0.05 defined statistically significant differences. The analysis was performed using a commercial statistical package of SPSS v24.0 (IBM Corp., Armonk, NY).

3 Results

3.1 Cellular Cementum in Premolar Tooth Roots (Fig. 1; Panels a1–a5: OIGF and Control Groups)

The mean cellular cementum thickness in the teeth root areas did not differ significantly

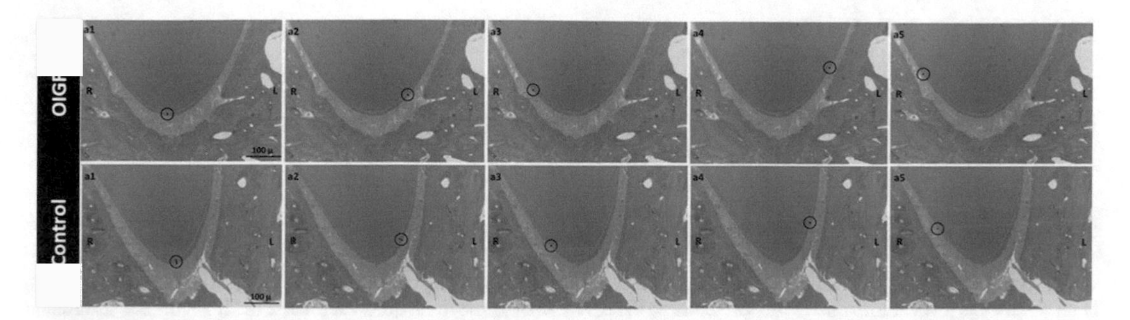

Fig. 1 Representative digital images of premolar root sections in a beagle dog. Thickness of cellular cementum layers is shown in the osteogenic-induced gingival fibroblasts (OIGF) group (upper row) and in the control group (lower row) at the root apex (a1) and at the following four measurement points (marked by black circles) referenced to the root apex: 500 μm off the apex left (distal side) (a2); 500 μ off the apex right (mesial side) (a3); 1000 μm off the apex left (a4); and 1000 μm off the apex right (a5). *R*, right side; *L*, left side. Each horizontal row in the figure corresponds to a separate tooth in the same dog. Magnification x2.52

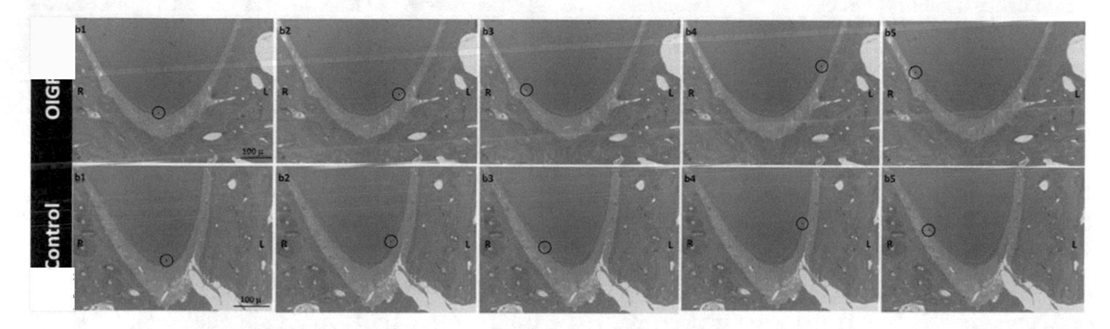

Fig. 2 Representative digital images of premolar root sections in a beagle dog. Thickness of acellular cementum layers is shown in the osteogenic-induced gingival fibroblasts (OIGF) group (upper row) and in the control group (lower row) at the root apex (b1) and at the following four measurement points (marked by black circles) referenced to the root apex: 500 μm off the apex left (distal side) (b2); 500 μm off the apex right (mesial side) (b3); 1000 μm off the apex left (b4); and 1000 μm off the apex right (b5). *R*, right side; *L*, left side. Each horizontal row in the figure corresponds to a separate tooth in the same dog. Magnification x2.52

between the OIGF and control groups at the five measurement points chosen in this study (Table 1). At the tooth root apex, cementum thickness amounted to 180.3 ± 44.0 μm and 186.9 ± 76.7 μm in the OIGF and control groups, respectively (Fig. 1; upper and lower panels a1). However, the mean cellular cementum thickness 500 μm off the apex bilaterally was about 1.5–2 times more in the OIGF group than in the control group, while a reverse trend was noticed 1000 μm off the apex, although the differences were insignificant due to large data scatter. Levene's test showed that data variability in the two groups was similar ($p > 0.05$), except for the measurement at 500 μm from the tooth root apex on the right side, where the assumption of variance homogeneity was violated.

3.2 Acellular Cementum in Premolar Tooth Roots (Fig. 2; Panels b1–b5: OIGF and Control Groups)

Likewise, the mean acellular cementum thickness in the teeth root areas did not differ significantly between the OIGF and control groups at the five measurement points (Table 2). At the tooth root apex, cementum thickness amounted to 35.3 ± 5.2 μm and 27.6 ± 5.6 μm in the OIGF and control groups, respectively (Fig. 2; upper

Table 1 Measurements of a cellular cementum thickness (μm) in different areas of teeth roots in the osteogenic-induced gingival fibroblasts (OIGF) and control groups

Cellular cementum	OIGF	*n*	Control	*n*	p-two sample	Mean Δ	Δ −95%CI	p-Levene's
Apical	180.3 ± 44.0	13	186.9 ± 76.7	13	0.94	−6.6 ± 88.4	−189.0; 175.9	0.37
Left 500 μm	87.7 ± 25.9	13	52.4 ± 11.4	11	0.25	35.3 ± 30.2	−27.1; 97.9	0.11
Left 1000 μm	67.3 ± 21.9	13	90.6 ± 42.3	12	0.62	23.3 ± 46.5	−119.6; 73.0	0.37
Right 500 μm	91.4 ± 24.1	13	47.9 ± 10.4	10	0.15	43.5 ± 29.1	−17.0; 104.0	0.04
Right 1000 μm	51.1 ± 11.1	13	96.1 ± 49.5	12	0.37	45.0 ± 48.9	−146.1; 56.1	0.08

Data are means ±SE. Left/right 500/1000 refers to the distance from the tooth apex on the left or right side; Δ, difference between OIGF and control groups; 95%CI, 95% lower and upper confidence intervals of the difference; Levene's test of variance homogeneity

Table 2 Measurements of a cellular cementum thickness (μm) in different areas of teeth roots in the osteogenic-induced gingival fibroblasts (OIGF) and control groups

Acellular cementum	OIGF	*n*	Control	*n*	p-two sample	Mean Δ	Δ −95%CI	p-Levene's
Apical	35.3 ± 5.2	13	27.6 ± 5.6	13	0.32	7.8 ± 7.6	−8.0; 23.6	0.84
Left 500 μm	22.7 ± 2.4	12	18.2 ± 2.6	11	0.22	4.5 ± 3.6	−2.9; 12.0	0.79
Left 1000 μm	20.9 ± 2.4	13	17.2 ± 2.1	11	0.27	3.7 ± 3.2	−3.0; 10.4	0.51
Right 500 μm	24.9 ± 2.9	12	19.7 ± 2.2	10	0.19	5.0 ± 3.8	−2.7; 13.2	0.33
Right 1000 μm	19.4 ± 2.1	13	17.7 ± 2.2	11	0.59	1.7 ± 3.0	−4.6; 8.0	0.87

Data are means ± SE. Left/right 500/1000 refers to the distance from the tooth apex on the left or right side; Δ, difference between OIGF and control groups; 95%CI, 95% lower and upper confidence intervals of the difference; Levene's test of variance homogeneity

and lower panels b1). In the other areas distant to the apex, differences in the acellular cementum thickness were rather meager and insignificant. Levene's test showed that data variability of all measurements was similar in the two groups ($p > 0.05$).

4 Discussion

This study investigated the effects of intraligamentary injection of osteogenic-induced gingival fibroblasts (OIGFs) on the tooth root cellular and acellular cementum thickness in beagle dogs subjected to orthodontic tooth resorption treatment. A histomorphometry evaluation was employed. According to the classic theory, the origin of osteoblasts and cementoblasts is the dental follicle proper, and perhaps also the perifollicular mesenchyme (Cho and Garant 2000; Ten Cate 1997). However, several reports indicate that cementoblasts could also originate from the Hertwig epithelial root sheath (Bosshardt and Selvig 1997; Thomas and Kollar 1988). Dogan et al. (2002) have shown that gingival fibroblasts have a potential to differentiate into cementoblast-like cells, whereas Mostafa et al. (2008) have also shown that gingival fibroblasts can differentiate into osteogenic cells to repair lost hard tissues. Our present results contradicted those findings as we found no appreciable effects of gingival fibroblasts on cementum thickness. Recently, Crossman et al. (2017) have conducted a study in beagle dogs on the effects of gingival fibroblasts on tooth root resorption evoked by orthodontic treatment. The results, akin to the present ones, show a lack of an enhancing effect of the fibroblasts on the thickness of cellular or acellular cementum.

In the present study, cellular and acellular cementum thicknesses were measured at the tooth root apex and at other four bilateral points distal to the apex. Although there was a general tendency for some increases in both cellular and acellular cementum in the OIGF group when compared with the control group, on average, no significant effects were noticed at any of the measurement points. A somehow stronger tendency for an increase in cellular than acellular cementum noticed in the OIGF group could be due to different mechanisms driving the formation of either type of cementum. Namely, cellular cementoblasts carry parathyroid hormone (PTH)-binding sites while acellular cementoblasts do not (Bosshardt 2005). Since cellular cementoblasts build up a thick mineralized layer in a short period of time, they respond to local factors in a much more pronounced manner than the acellular ones. Thus, acellular cementum needs more time to shape into a final mold. In this context, it is worth noting that the current study ended after 28 days. It cannot be excluded that a longer observation time, an increase in cell density, or multiple injections of OIFG through the periodontal ligament near the root apex would yield a more distinct stimulatory effect on the cementum growth. Likewise, future studies might consider a long-term cell labeling to provide a cause-effect assessment of the intraligamentary injection of OIGF regarding the increase in cementum thickness in either regeneration or repair of orthodontic-induced tooth root resorption. A study on a larger sample of cases would also be required to decisively settle the issue of the influence of gingival fibroblasts on tooth root cementum.

In conclusion, a single intraligamentary injection of osteogenic-induced gingival fibroblasts (OIGF) failed to appreciably increase the tooth root cementum thickness in beagle dogs with modeled orthodontic tooth root resorption. Thus, OIGF is unlikely to prevent orthodontic-induced tooth root resorption.

Conflicts of Interest The authors declare no conflicts of interest in relation to this study.

Ethical Approval All applicable international, national, and institutional guidelines for the care and use of animals were followed. Ethical approval for this research was received from the University of Alberta Animal Research Ethics Committee. The experiments complied with the guidelines of Animal Research: Reporting of In Vivo Experiments (ARRIVE) and were carried out in accord with the UK Animals Act (Scientific Procedures) 1986 and with the EU Directive 2010/63/EU for animal experiments.

References

Bosshardt DD (2005) Are cementoblasts a subpopulation of osteoblasts or a unique phenotype? J Dent Res 84:390–406

Bosshardt DD, Selvig KA (1997) Dental cementum: the dynamic tissue covering of the root. Periodontology 2000 13:41–75

Brezniak N, Wasserstein A (2002) Orthodontically induced inflammatory root resorption. Part I: the basic science aspects. Angle Orthod 72:175–179

Chan E, Darendeliler MA (2006) Physical properties of root cementum: part 7. Extent of root resorption under areas of compression and tension. Am J Orthod Dentofac Orthop 129:504–510

Cho MI, Garant PR (2000) Development and general structure of the periodontium. Periodontology 24:9–27

Crossman J, Hassan AH, Saleem A, Felemban N Aldaghreer S, Fawzi E, Farid M, Abdel-Ghaffar K, Gargoum A, El-Bialy T (2017) Effect of gingival fibroblasts and ultrasound on dogs' root resorption during orthodontic treatment. J Orthod Sci 6:28–35

de Vries TJ, Schoenmaker T, Wattanaroonwong N, van den Hoonaard M, Nieuwenhuijse A, Beertsen W, Everts V (2006) Gingival fibroblasts are better at inhibiting osteoclast formation than periodontal ligament fibroblasts. J Cell Biochem 98:370–382

Dogan A, Ozdemir A, Kubar A, Oygur T (2002) Assessment of periodontal healing by seeding of fibroblast-like cells derived from regenerated periodontal ligament in artificial furcation defects in a dog: a pilot study. Tissue Eng 8:273–282

Foster BL (2012) Methods for studying tooth root cementum by light microscopy. Int J Oral Sci 4:119–128

Grzesik WJ, Narayanan AS (2002) Cementum and periodontal wound healing and regeneration. Crit Rev Oral Biol Med 13:474–484

Harry MR, Sims MR (1982) Root resorption in bicuspid intrusion. A scanning electron microscope study. Angle Orthod 52:235–258

Jager A, Kunert D, Friesen T, Zhang D, Lossdorfer S, Gotz W (2008) Cellular and extracellular factors in early root resorption repair in the rat. Eur J Orthod 30:336–345

Killiany DM (1999) Root resorption caused by orthodontic treatment: an evidence-based review of literature. Semin Orthod 5:128–133

Mirabella AD, Artun J (1995) Prevalence and severity of apical root resorption of maxillary anterior teeth in adult orthodontic patients. Eur J Orthod 17:93–99

Mohammadi M, Shokrgozar MA, Mofid R (2007) Culture of human gingival fibroblasts on a biodegradable scaffold and evaluation of its effect on attached gingiva: a randomized, controlled pilot study. J Periodontol 78:1897–1903

Mostafa N, Scott P, Dederich D, Doschak M, El-Bialy T (2008) Low intensity pulsed ultrasound stimulates osteogenic differentiation of human gingival fibroblasts. Proc Ann Mtg Can Acoustics Assoc 36:34–35

Motokawa M, Sasamoto T, Kaku M, Kawata T, Matsuda Y, Terao A, Tanne K (2012) Association between root resorption incident to orthodontic treatment and treatment factors. Eur J Orthod 34:350–356

Remington DN, Joondeph DR, Artun J, Riedel RA, Chapko MK (1989) Long-term evaluation of root resorption occurring during orthodontic treatment. Am J Orthod Dentofac Orthop 96:43–46

Ten Cate AR (1997) The development of the periodontium-a largely ectomesenchymally derived unit. Periodontology 2000(13):9–19

Thomas H, Kollar E (1988) Tissue interactions in normal murine root development. In: Davidovitch Z (ed) The biological mechanisms of tooth eruption and root resorption. EBSCO Media, Birmingham, pp 145–151

Tobita M, Uysal AC, Ogawa R, Hyakusoku H, Mizuno H (2008) Periodontal tissue regeneration with adipose-derived stem cells. Tissue Eng Part A 14:945–953

Yamamoto T, Hasegawa T, Yamamoto T, Hongo H, Amizuka N (2016) Histology of human cementum: its structure, function, and development. Jpn Dent Sci Rev 52:63–74

Zeichner-David M (2006) Regeneration of periodontal tissues: cementogenesis revisited. Periodontology 2000(41):196–217

Adv Exp Med Biol - Clinical and Experimental Biomedicine (2021) 11: 115–123
https://doi.org/10.1007/5584_2020_555

Published online: 22 June 2020

In Silico Evaluation of Treatment of Periprosthetic Fractures in Elderly Patients After Hip Arthroplasty

Jacek Lorkowski, Renata Wilk, and Mieczyslaw Pokorski

Abstract

The aim of this study was to investigate the soundness of in silico finite element model (FEM) in the assessment of strain in the femur and in the components fixing the periprosthetic fracture in elderly patients after hip arthroplasty. From a group of 55 patients, aged 27–95, treated due to fractures after hip replacement in 2012–2018, 18 patients were separated out, aged over 85, out of whom 7 had type C fractures, according to the Vancouver classification. These seven patients formed the study group. The fractures were stabilized with a locking compression plate system and wire loops or by replacement of the endoprosthesis stem. The FEM was performed by processing radiological images of the femur, considering the stabilization type and osteoporotic bone characteristics. Each patient's FEM was counter compared to virtual in silico control showing a non-osteoporotic bone structure. We found that the strain was distinctly greater at the bone-implant interface after surgical stabilization with a multi-hole plate and cerclage wire loops in osteoporotic periprosthetic fractures when compared to the virtual non-osteoporotic bone. We conclude that the in silico model enables the assessment of strain distribution at the bone-implant interface, which helps identify the biomechanical incongruity of traditional bone stabilization methods in patients with osteoporotic bones.

Keywords

Elderly patients · Endoprosthesis · Finite element model · Hip arthroplasty · In silico · Periprosthetic fracture

J. Lorkowski (✉)
Department of Orthopedics and Traumatology, Central Clinical Hospital of Ministry of Interior, Warsaw, Poland
e-mail: jacek.lorkowski@gmail.com

R. Wilk
Department of Anatomy, Health Science Department, Medical University of Silesia, Katowice, Poland

Hope Medical Institute, Newport News, VA, USA

M. Pokorski
Institute of Health Sciences, Opole Medical School, Opole, Poland

1 Introduction

Hip arthroplasty is a frequent procedure to improve quality of life, extend lifespan, and maintain mobility in osteoarthritis (Marshall et al. 2017). In the USA, more than 300,000 such procedures are performed annually, mostly in the elderly people. Technological improvements in implant production and advances in endoprosthesis implantation cause that treatments are carried out in increasingly elderly patients with weak bone structure (Caruso et al. 2018). While the procedure of implant placement usually ends successfully, complications such as periprosthetic fractures are on the rise, reaching 1.7–4.1% of all arthroplasty procedures (Lee et al. 2018; Marshall et al. 2017). The fractures appear more often after revision arthroplasty than after primary endoprosthesis placement, 6% vs. 1%, respectively. The most frequent cause of periprosthetic fracture is a fall on the flat surface caused by slipping or tripping. However, there also are fractures without previous injury, which points to bone weakness that raises concern in the elderly over 85 years of age who, additionally, may have various comorbidities (Chou and Davis 2017; Marsland and Mears 2012). Female gender, osteoporosis with bone mass loss, rheumatoid arthritis, and osteolytic lesions, particularly in bone areas exposed to load, predispose to fractures (Caruso et al. 2018). According to the Swedish National Hip Arthroplasty Registry, the incidence of periprosthetic fractures is second only to bone infections as complications after arthroplasty, raising socioeconomic issues due to morbidity and mortality (Karrholm et al. 2016). Several classification systems have been develop to assess periprosthetic fractures, which have a varying ability to characterize the fracture in radiographs, the condition of the tissue surrounding bone implant, and the implant fixation (Tower and Beals 1999; Cooke and Newman 1988; Johansson et al. 1981). Currently, Vancouver classification is commonly used. It enables the assessment of fracture size, prosthesis required, and the selection of the best treatment method (Duncan and Haddad 2014).

The Vancouver classification consists of the following facture types:

- Type A, fracture passing through the greater and lesser trochanters;
- Type B, fracture along the length of endoprosthesis, passing through the medial margin of the femur, which is further subdivided into:
  - Type B1, with preserved implant stability and good quality of bone matrix around the implant;
  - B2, with loose endoprosthesis but good quality of bone structure;
  - B3, with loss of stability of endoprosthesis and bone of deficient structure;
- Type C, fracture extending below the endoprosthesis (Marshall et al. 2017).

The assessment of implant fixation in type B and C fractures is often subjectively made by a surgeon who performs anastomosis (Huang et al. 2018), which may have a bearing on a greater number of reoperations in these categories. Type C fractures create the largest clinical difficulty since they hamper balance and locomotor stability, often resulting in repeat fractures (Pavone et al. 2019; Randelli et al. 2018). The current standard is to schedule surgery, based on the evaluation of the patient's condition supported by radiographs. In advanced age, extensive procedures take toll on the patient, so that surgical decisions should be weight in with caution. In such situations, the finite element method (FEM), based on a rapid 2D modeling, seems of potential applicability, to assess the distribution of strain occurring in the implanted bone. Therefore, the main purpose of this study was to investigate the soundness of an in silico FEM model for the assessment of strain in the femur and in the components fixing the periprosthetic type C fractures in elderly patients after hip arthroplasty.

2 Methods

In this retrospective study, 18 elderly patients, aged over 85, were separated out of a group of 55 patients, aged 27–95, treated due to femur fractures after hip replacement surgery performed between 2012 and 2018. Seven out of the 18 patients had type C fractures, according to the Vancouver classification. These seven patients formed the study group. The fractures were stabilized with a locking compression plate system and wire loops or by replacement of the endoprosthesis stem. Out of the remaining 18 patients, 5 had type B1, 3 had type B2, and another 3 had type B3 fractures. Type B fractures, usually occurring in bones having better structural quality, were not considered in this study.

Treatment outcome in type C fractures was evaluated 1 year after surgery using the Harris Hip Score (HHS) for the assessment of pain and joint mobility. The mean HHS score for this group was 71 ± 4.4 (SD) out of 100 points possible to achieve, which is considered a fair result (Harris 1969). In silico rapid modeling of fracture stabilization was based on patients' X-ray images of the femur obtained in the process of diagnosis and treatment. The method evaluates the strain arising in solid-state elements, which the bone belongs to, in the area initially divided into a specific number of simple finite elements that connect to each other at certain points in the periphery, called nodes (Lorkowski et al. 2018). After determining the distribution of the physical quantity inside finite elements, i.e., shape functions of nodes, calculations of differential equations for nodes are performed, while in the remaining parts of finite elements, the equation solutions are approximated. Then, boundary conditions are introduced into the system of equations, and the values of the physical quantities of interest in the nodes of interest are obtained. In this study, the degree and type of displacement in the nodes were determined, which allowed for the calculation of bone strain values.

Patients' X-ray images were used to make a geometric model of the femur with the implant and anastomosis used to stabilize the periprosthetic fracture. The model simulated fixation used for a fracture and stabilization depending on the bone parameters in elderly patients. Each osteoporotic patient's FEM was counter compared to virtual in silico control, containing the assumed healthy-like bone structure of a young adult with the corresponding distribution of elemental strains. The images were subjected to in silico modeling combining bone dynamics and FEM simulation, which allowed to visualize specific changes occurring in bones with different than normal characteristics, e.g., due to osteoporosis but also caused by the type fixation and by the endoprosthesis itself. The FEM model was created by automatically converting X-ray images considered as a bitmap composed of 256 grayscale elements in the computer program CT2FEM 1.0. The program assigns the appropriate FEM value to each modulus of the grayscale element. Next, the created model was loaded into Ansys Discovery Design software (Canonsburg, PA) to divide the finite elements into groups representing different tissues, such as compact or spongy bone, and the materials the endoprosthesis and fracture fixation components were made of.

Bone strain can be described by Young's coefficient and Poisson's coefficient for elasticity of solids. Young's modulus is the ratio of tension arising in the material to the elastic deformation caused under the influence of acting force. It varies depending on the tissue being analyzed. In our case, 4 values were important, representing the examined tissues and the metal used for fixation: 5 GPa for compact bone, 3 GPa for spongy bone, 300 MPa for connective tissue and marrow cavity, and 210 GPa for fixation metal. In case of advanced osteoporosis in the elderly, the above standard values representing Young's modulus of bone elements were halved. Poisson's coefficient, on the other side, is the ratio between transverse strain to the corresponding axial strain. It was determined as 0.32 for all the tissues. Fragments of the FEM model representing different regions could be turned off or on as separate subgroups for the analysis. All fragments of a given type of tissue were treated in the same way (Wirtz et al. 2000).

3 Results

Final analysis of in silico modeling included the following types of stabilization of type C fractures on osteoporotic patients:

1. Multi-hole plate, three cerclage wire loops, and four screws in the distal segment.
2. Multi-hole plate, three cerclage wire loops, and five screws in the distal segment.

The screws in the distal segment were passing through the whole bone, i.e., across the opposite layers of cortical bone. Additionally, at the level of endoprosthesis in the proximal segment, there were three screws, passing through only one cortical bone layer, and cerclage wire loops in both types of multi-hole plate stabilization above outlined. The results of FEM analysis obtained for either type of stabilization is exemplified below.

Re 1 Peri-implant strain in an elderly patient with advanced osteoporosis and a femur fracture stabilized using a four-screw multi-hole plate and three cerclage wire loops was 792.215 MPa at the fracture level (Fig. 1a). For comparison, the strain was reduced to 738.189 MPa, i.e., by 6.8% in the corresponding virtual model of a non-osteoporotic bone (Fig. 1b). Likewise, there were reductions in strain at the levels of the remaining three screws in the hypothetical young bone. Reductions in strain were also noticed both above and below the anastomosis. The largest strain in the patient's in silico model, double that present at the fracture site, appeared below the downmost screw, where there also was the largest strain reduction in the virtual model of young bone, exceeding 100 MPa.

Re 2 Peri-implant strain in an elderly patient with advanced osteoporosis and a periprosthetic femur fracture stabilized with a five-screw multi-hole plate and three cerclage wire loops was 789.956 MPa at the fracture level (Fig. 2a). The strain was reduced to 737.180 MPa, i.e., by about 7%, in the corresponding virtual model of a control non-osteoporotic bone (Fig. 2b). Akin to the four-screw stabilization model, the strain was increasing toward the distal bone segments in both true osteoporotic and virtual non-osteoporotic bone

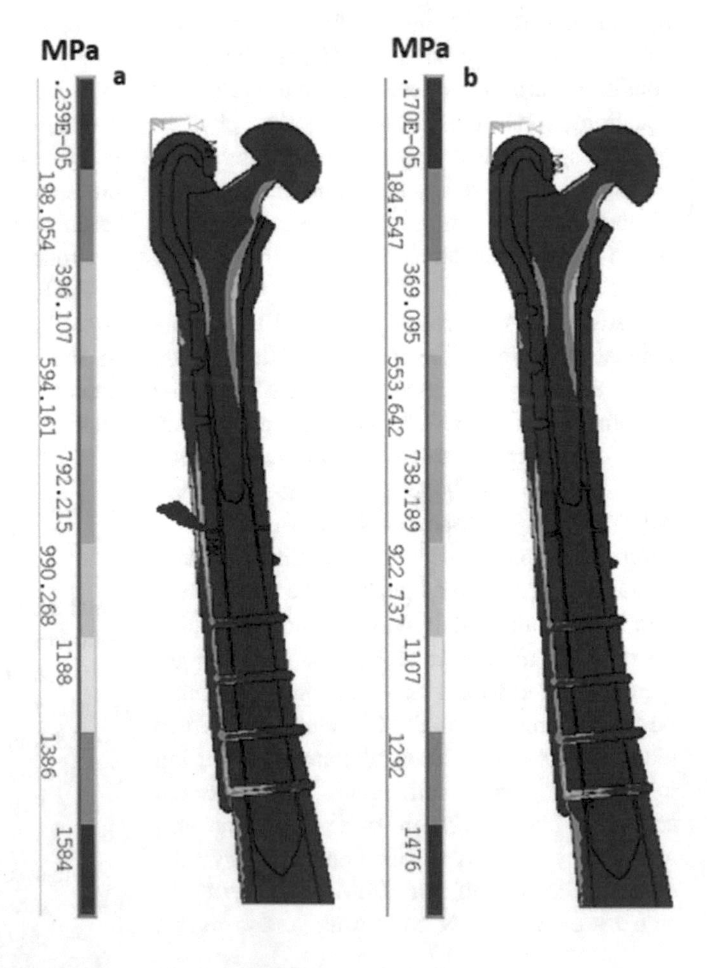

Fig. 1 Frontal plane image. In silico finite element method (FEM) analysis of distribution of biomechanical strain in periprosthetic type C femur fracture stabilized with a multi-hole plate and four screws distally. Additional three shorter screws, passing through only one cortical bone layer, and three-wire couplings are visible in the proximal section: (**a**) real image of osteoporotic bone in an elderly patient, (**b**) virtual model corresponding to the same but non-osteoporotic, hypothetically healthy-like bone; red mark in "a" marks the fracture gap

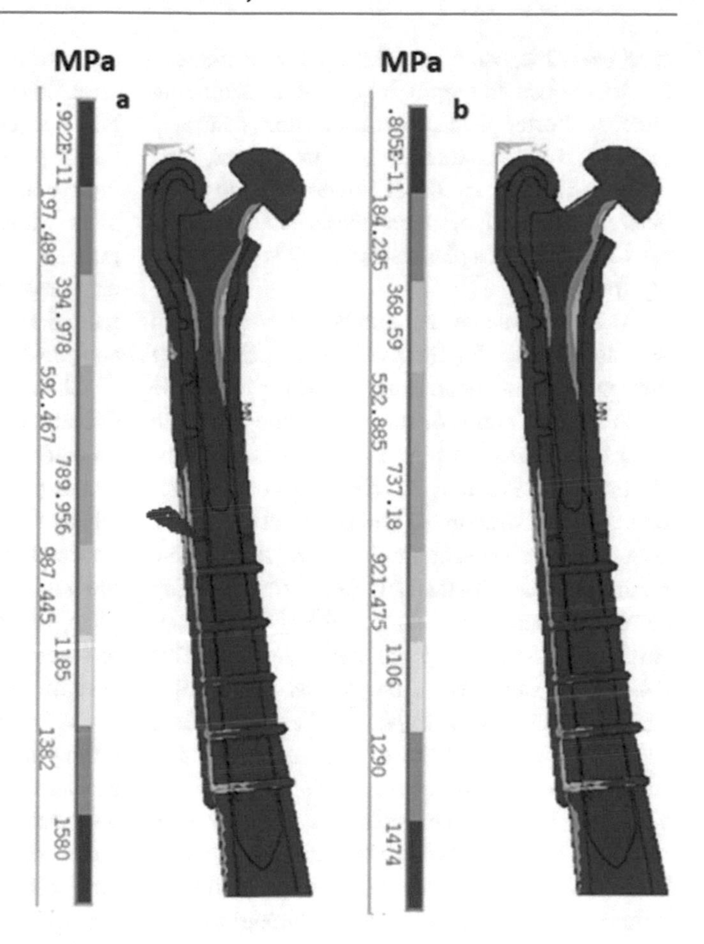

Fig. 2 Frontal plane image. In silico finite element method (FEM) analysis of distribution of biomechanical strain in periprosthetic type C femur fracture stabilized with a multi-hole plate and five screws distally. Additional three screws, passing through only one cortical bone layer, and three-wire couplings are visible in the proximal section: (**a**) real image of osteoporotic bone in an elderly patient, (**b**) virtual model corresponding to the same but non-osteoporotic, hypothetically healthy-like bone; red mark in "a" marks the fracture gap

models, being correspondingly lower by about 10% in the latter bone model. The strain amounted to 1580.000 and 1474.000 MPa, i.e., by 6.7% in the downmost bone segment, respectively (Fig. 2b).

Strain evaluation was also made after stabilization of a prosthetic femur fracture using a multiplate four screws passing only through one cortical bone layer below endoprosthesis as seen in the proximal sections of the presented images. Differences between the true osteoporotic fractures and their corresponding virtual vis-a-vis were here less expressive. The bone strain was comparable in both osteoporotic and non-osteoporotic models above the fracture, in a range of 592 MPa and 599 MPa. The maximum strain was noticed at the fracture level, amounting to 2,309 MPa in osteoporotic and 2,397 MPa in non-osteoporotic segments, an increase by about 4%. Stabilization of a femur fracture using a multi-hole plate and five screws passing through one cortical layer provided comparable strain values (data not shown). Thus, mounting the plate only to the underlying cortical bone layer blurs the difference in strain between the osteoporotic and non-osteoporotic conditions and shows a fourfold increase in strain at the fracture level.

4 Discussion

Periprosthetic fractures are a serious complication in patients after hip arthroplasty. Especially, type C fractures that appear below the prosthesis shaft are challenging to treat. Typically, treatment includes dynamic compression plates, plates and cerclage wires, and non-locking plates. It is essential to choose the right length of a stabilizing plate. A shorter plate allows for a shorter surgical incision, reduces trauma to soft tissues, and shortens the surgery time. However, bone strain between the end of a prosthesis shaft and the proximal edge of a plate is high, risking a secondary fracture.

Another issue is the number of screws and wire loops used for the plate mount. Increasing their number and extent may cause unpredictable shifts in bone strain or raise substitute strains in other bone parts, leading to dysfunctional bone elasticity and eventually to prosthesis destabilization or repeat fractures (Lorkowski et al. 2014). Based on the present findings, we submit that mounting a metal plate using four screws, passing across the bone, with three cerclage wire loops suffices to achieve the optimum outcome. This conclusion stems from the results of in silico FEM modeling which showed that an increase in the number of screws to five caused a rather superfluous drop of about 4–5% in the strain along the femur bone, including the level of a multi-hole plate when compared to four screws. The assessment we made mostly concerned bone biomechanics. It did not include a biological context, for instance, peri-screw or bone itself infections or other bone pathologic conditions. It did include however osteoporosis, all too often occurring in elderly patients. Further, rapid comparing of the true patient's osteoporotic bone with its virtual non-osteoporotic, healthy-like counterpart enabled the assessment of changes in the strain distribution pattern caused by osteoporosis. We noticed that osteoporosis enhanced the bone strain around the multi-plate stabilator of a femur fracture by about 7%, irrespective of the use of four or five screws. Nonetheless, the enhanced strain in periprosthetic fracture due to osteoporosis may be viewed as another argument to use fewer rather than more screws to maintain adequate osteoporotic bone elasticity.

When stabilizing a fracture, it is necessary to use cerclage wire loops above the prosthesis. Some surgeons also use additional shorter screws that pass through one cortical layer, reaching down to the level of an endoprosthesis (O'Toole et al. 2006). The in silico model of this study failed to confirm that the use of cerclage wire loops prevents the occurrence of a relatively high bone strain at the level of a multi-hole plate. Moreover, the use of plate screws passing through only one cortical layer at the level of endoprosthesis did not lead to a significant strain reduction along the entire length of bone subjected to anastomosis.

Choosing a correct method of stabilizing type C fractures is a matter of uncertainty in therapeutic decision-making. The questions arise to what extent plate osteosynthesis should be used and whether plate stabilization of a fracture should be combined with the revision prosthesis. An obvious disadvantage of stabilization with a plate, particularly in the elderly, is a longer time required for mobilization of a limb operated on. In case of revision arthroplasty, physical mobilization is possible much earlier. However, arthroplasty, particularly with additional plate stabilization, is a much more extensive and thus aggravating procedure. In principle, all types of anastomosis share similar doubts. In patients with periprosthetic fractures, treated in a classic manner with open reduction of bone displacement and internal fixation (ORIF), the percentage of failures reaches 33% in all fracture types (Pavone et al. 2019). Zuurmond et al. (2010) have shown that 12 (55%) of 21 patients with type C fractures, who had a standard anastomosis with a non-locking plate or plates with cerclage wire loops, required reoperation. Better results have been noticed using locking plates in the less invasive stabilization system (LISS) for fracture fixation in another study, in which 11 of 12 patients have a post-surgery course free of complications (O'Toole et al. 2006). In the remaining patient, the plate broke down and required a replacement into a longer one. In another study, open reduction of bone displacement and double-plate fixation resulted in good outcome, with anastomosis union, in six of nine patients with type C periprosthetic femoral fractures (Lee et al. 2018). The three remaining patients had poor outcomes with nonunion, due mainly to osteoporosis. In two of those three patients (80 and 82 years of age), a secondary femur fracture took place below the anastomosis 2 and 4 weeks after fixation with double plates and wires. The two patients underwent revision surgery, using a double-plate anastomosis covering a longer length of a femur, with good long-term outcomes. In the third patient, in whom the plate failed to anastomose resulting in a new fracture, good outcome also was present 24 months after replacement for a double-plate anastomosis.

Patients with type C fractures supplied with classic ORIF stabilization have often recurrent fractures that are far more difficult to treat (Randelli et al. 2018). Another issue is revision prosthesis replacement, if required, using a compression plate to stabilize the periprosthetic fracture. We have previously described an elderly patient with perioperative fracture in whom the distribution of strain occurring during fracture stabilization was at the bone-anastomosis boundary greater compared to the average control. A high level of strain appeared linked to the excessive number of cerclage wire loops used (Lorkowski et al. 2014). These results give a consistent impression that in elderly patients with osteoporosis, treated with classic anastomosis, a check on the strain distribution over the bone should be an indispensable clinical intervention to achieve the best post-surgery outcome and to minimize a risk of secondary fractures and reoperations. To this end, FEM modeling enables the selection of the most appropriate method of anastomosis while scheduling surgery. A 2D version of FEM provides prompt results, approximately in 1–2 h, which seems the optimum time to fix a fracture from the biomechanical standpoint, as opposed to 3D modeling that due to structural complexity requires 20- to 30-fold longer timeframe.

In patients with periprosthetic fractures, X-rays are part of routine examination, as opposed to less used computed tomography. At present, X-rays are the best source for in silico FEM modeling. The method contains several software-driven simplifications that tend to unify various tissues examined. Therefore, a model contains approximations and rounding of details of bony structure, but the evaluation appears sufficient to illustrate changes taking place in the bone strain after application of different types of anastomosis. Further, the method appears well-suited to appraise the level of osteoporosis, particularly in the elderly patient and spare additional complications and suffering related to surgical revisions, prosthesis exchanges, or repeat fractures.

FEM modeling, based on the solid-state biomechanics, was originally created to aid arthroplasty and orthopedic implants. Recently, the method has become of increasing interest due to potential diagnostic and therapeutic utility also in other medical fields, with some adaptations to fluid mechanics. The method has been studied, inter alia, for the prediction of an extent of post-traumatic lung injuries, which includes the evaluation of the accompanying lung inflammatory components (Danelson and Stitzel 2015; Lorkowski et al. 2015). Other novel applications of FEM extend to the evaluation of the biomechanical properties of the eustachian tube that connects the middle ear to the nasopharynx, which has to do with cartilage stiffness and mucosal adhesion in otitis media (Malik et al. 2016). Modeling of coiling deployment during endovascular interventions for intracranial aneurysms (Damiano et al. 2019) or the evaluation of a rupture risk of abdominal aortic aneurism (Siika et al. 2019; Soto et al. 2018) are other examples of this apparently burgeoning field of medicine. Biomechanical fingerprinting of lung tissue perfusion changes underlying inflammatory disorders seems an intriguing possibility to enhance the diagnostics of SARS-CoV-2 pneumonitis and the like.

In conclusion, using the finite element modeling, we found that stabilization of type C periprosthetic fractures in the elderly patients is bound to enhance femoral strain at the bone-implant interface, due likely to osteoporotic bone changes. The strain is distinctly greater compared to that present in the young bone counterpart. In silico modeling enables the assessment of bone strain arising in periprosthetic fractures after hip arthroplasty, which helps identify the biomechanical incongruity of traditional bone stabilization methods.

Conflicts of Interest The authors declare that they have no conflicts of interest in relation to this article.

Ethical Approval All procedures performed in studies involving human participants were in accordance with the ethical standards of the institutional and/or national research committee and with the 1964 Helsinki Declaration and its later amendments or comparable ethical standards. Due to a retrospective nature of the material used in this study, a requirement to obtain approval from a local ethics committee as well as informed consent from individual participants were waived.

References

Caruso G, Milani L, Marko T, Lorusso V, Andreotti M, Massari L (2018) Surgical treatment of periprosthetic femoral fractures: a retrospective study with functional and radiological outcomes from 2010 to 2016. Eur J Orthop Surg Traumatol 28:931–938

Chou DTS, Davis B (2017) Trochanteric femoral fracture around a Birmingham hip resurfacing prosthesis: a case report and review of the literature. JBJS Case Connect 7(1):e7

Cooke PH, Newman JH (1988) Fractures of the femur in relation to cemented hip prostheses. J Bone Joint Surg Br 70(3):386–389

Damiano RJ, Tutino VM, Lamooki SR, Paliwal N, Dargush GF, Davies JM, Siddiqui AH, Meng H (2019) Improving accuracy for finite element modeling of endovascular coiling of intracranial aneurysm. PLoS One 14(12):e0226421

Danelson KA, Stitzel JD (2015) Finite element model prediction of pulmonary contusion in vehicle-to-vehicle simulations of real-world crashes. Traffic Inj Prev 16(6):627–636

Duncan CP, Haddad FS (2014) The unified classification system (UCS): improving our understanding of periprosthetic fractures. Bone Joint J 96(6):713–716

Harris WH (1969) Traumatic arthritis of the hip after dislocation and acetabular fractures: treatment by mold arthroplasty. An end-result study using a new method of result evaluation. J Bone Joint Surg Am 51:737–755

Huang JF, Jiang XJ, Shen JJ, Zhong Y, Tong PJ, Fan XH (2018) Modification of the unified classification system for periprosthetic femoral fractures after hip arthroplasty. J Orthop Sci 23(6):982–986

Johansson JE, McBroom R, Barrington TW, Hunter GA (1981) Fracture of the ipsilateral femur in patients with total hip replacement. J Bone Joint Surg Am 63 (9):1435–1442

Karrholm J, Lindahl K, Malchau H, Mohaddes M, Nemes S, Rogmark C, Rolfson O (2016) The Swedish hip arthroplasty register. Annual report 2016. https://registercentrum.blob.core.windows.net/shpr/r/Annual-Report-2016-B1eWEH-mHM.pdf. Accessed on 12 Oct 2018

Lee J, Kim T, Kim T (2018) Treatment of periprosthetic femoral fractures following hip arthroplasty. Hip Pelvis 30(2):78–85

Lorkowski J, Mrzygłód MW, Kotela A, Kotela I (2014) Application of rapid computer modeling in the analysis of the stabilization method in intraoperative femoral bone shaft fracture during revision hip arthroplasty – a case report. Pol Orthop Traumatol 79:138–144

Lorkowski J, Mrzygłód M, Grzegorowska O (2015) Finite elements modeling in diagnostics of small closed pneumothorax. Adv Exp Med Biol 866:7–13

Lorkowski J, Grzegorowska O, Kozień MS, Kotela I (2018) Effects of breast and prostate cancer metastases on lumbar spine biomechanics: rapid *in silico* evaluation. Exp Med Biol 1096:31–39

Malik JE, Swarts JD, Ghadiali SN (2016) Multi-scale finite element modeling of Eustachian tube function: influence of mucosal adhesion. Int J Numer Method Biomed Eng 32(12):e02776

Marshall RA, Weaver MJ, Sodickson A, Khurana B (2017) Periprosthetic femoral fractures in the emergency department: what the orthopedic surgeon wants to know. Radiographics 37:1202–1217

Marsland D, Mears SC (2012) A review of periprosthetic femoral fractures associated with total hip arthroplasty. Geriatr Orthop Surg Rehabil 3(3):107–120

O'Toole RV, Gobezie R, Hwang R, Chandler AR, Smith RM, Estok DM, Vrahas MS (2006) Low complication rate of LISS for femur fractures adjacent to stable hip or knee arthroplasty. Clin Orthop Relat Res 450:203–210

Pavone V, de Cristo C, Di Stefano A, Costarella L, Testa G, Sessa G (2019) Periprosthetic femoral fractures after total hip arthroplasty: an algorithm of treatment. Injury 50(Suppl 2):S45–S51

Randelli F, Pace F, Priano D, Giai Via A, Randelli P (2018) Re-fractures after periprosthetic femoral fracture: a difficult to treat growing evidence. Injury 49 (Suppl 3):S43–S47

Siika A, Lindquist Liljeqvist M, Zommorodi S, Nilsson O, Andersson P, Gasser TC, Roy J, Hultgren R (2019) A large proportion of patients with small ruptured abdominal aortic aneurysms are women and have chronic obstructive pulmonary disease. PLoS One 14 (5):e0216558

Soto B, Vila L, Dilmé J, Escudero JR, Bellmunt S, Camacho M (2018) Finite element analysis in symptomatic and asymptomatic abdominal aortic aneurysms for aortic disease risk stratification. Int Angiol 37 (6):479–485

Tower SS, Beals RK (1999) Fractures of the femur after hip replacement: the Oregon experience. Orthop Clin North Am 30(2):235–247

Wirtz DC, Schiffers N, Pandorf T, Radermacher K, Weichert D, Forst R (2000) Critical evaluation of known bone material properties to realize anisotropic FE-simulation of the proximal femur. J Biomech 33 (10):1325–1330

Zuurmond RG, van Wijhe W, van Raay JJ, Bulstra SK (2010) High incidence of complications and poor clinical outcome in the operative treatment of periprosthetic femoral fractures: an analysis of 71 cases. Injury 41 (6):725–733

Adv Exp Med Biol - Clinical and Experimental Biomedicine (2021) 11: 125–131
https://doi.org/10.1007/5584_2020_569

Published online: 22 July 2020

Appraisal of Burden of Caregivers to Chronically Rehabilitated Patients with Spinal Cord Injuries in a Tertiary Neurological Center in Nepal

Sunil Munakomi, Arzu Bajracharya, Suja Gurung, Marwin Dewan, Narendra Prasad Joshi, Avinash Mishra, Kusum Dhamala, Sangam Shrestha, Kanchan Bharati, and Giovanni Grasso

Abstract

The care of a patient with a spinal cord injury is part of healthcare systems. It causes a substantial physical and emotional drain on the caretakers who often are in short supply and thus may lack adequate training, preparation, and support. Long hours of assisting a chronically handicapped patient with activities of daily living and exercises decrease the rehabilitator's quality of life and take a psychological toll that increases a risk of burnout syndrome. The present study found a significant caregiving burden among care providers of chronically dependent patients with spinal cord injuries. Additionally, financial drain escalates the issue in this rather neglected health and quality of life aspect concerning caregivers. For the situation to improve, there must be a paradigm shift in care taking toward the motivative patient's participation in the rehabilitative process. Provisions for social support and educational programs focusing on the patients and their families need to be reappraised.

Keywords

Caregiver · Disability care · Disabled patient · Emotional drain · Healthcare · Quality of life · Rehabilitation · Spinal cord injury

S. Munakomi (✉), A. Bajracharya, S. Gurung, and K. Dhamala
Neurosurgery, Nobel Medical College and Teaching hospital, Biratnagar, Nepal
e-mail: sunilmunakomi@gmail.com

M. Dewan, N. P. Joshi, and A. Mishra
Physiotherapy Unit, Nobel Medical College and Teaching hospital, Biratnagar, Nepal

S. Shrestha
Pediatrics, Koshi Zonal Hospital, Biratnagar, Nepal

K. Bharati
Neurosurgery, Purbanchal University, Chitwan, Nepal

G. Grasso
Neurosurgical Clinic, Department of Biomedicine, Neurosciences and Advanced Diagnostics University of Palermo, Palermo, Italy

1 Introduction

Spinal cord injury (SCI) accounts for partial or total, temporary or permanent loss of motor, sensory, and autonomic functions (Ditunno et al. 1994). The global incidence of SCI is, roughly estimated, 236 to 1009 new cases annually per million persons (Cripps et al. 2011), and it is

12–61 cases per million in Asia alone (Burns and O'Connell 2012). Thus, SCI remains one of the major public health concerns worldwide. SCI not only encompasses physical limitations and psychological difficulties to the patient, but it also hampers in a multitude of ways the life of a patient's family: in often cases the primary caregivers. There also is an undeniable prospect that the economic cost relating to the management of SCI is going to be high. A conservative estimate of a mean lifetime cost of about 1300 new SCI cases arising annually in the UK varies from £1.1 to £1.9 million, depending on the severity of spinal injury, as assessed in 2016 prices (McDaid et al. 2019). In the context of low-income nations like Nepal, with a gross per capita income of mere 1025 dollars or less (World Bank 2019), people are bound to face substantial economic barriers. Strict limits in annual health resources, including neurological disorders that often require surgery followed by chronic care and rehabilitation, have crippling financial effects on families. The post-surgery rehabilitation phase, a pivotal need for patients with SCI, is hardly possible in the hospital or ambulatory care setting; thus, patients are usually discharged to be cared for by family members. The Kathmandu district has one of the highest prevalence of physically disabled people (based on the 2011 census) (DATAVIZ 2016). Yet rehabilitation centers dedicated to managing this subset of patients are alarmingly lacking, with only a single existing facility to serve neurorehabilitation in the country. This provides a ripple damage as the whole burden of management is shifted to caregivers who are abruptly forced into a position of facing the interplay between family member's disease, handicap, and the lacking resources. Caregivers must adapt to balance their time, energy, and focus between their job, family, and the caretaking role, the problems that take a heavy psychosocial toll, all of which having withered emotional effects (Middleton et al. 2014). Ironically, the attention to the burden faced by the caregiver is an underreported and neglected contextual matter. Therefore, in this study we address the issue by determining the quantify of various domains of burden exhibited by caregivers of patients with SCI, based on the Caregiver Burden Inventory (CBD) scale, with the aim to bring new insights into the strategy of chronic management of severely handicapped neurological patients.

2 Methods

This is a questionnaire-based study performed in 71 primary caregivers of SCI patients, in the disabled individual's home, who were unable to self-support activities of daily living. The questionnaire was the Caregiver Burden Inventory (CBD) (Marvardi et al. 2005). It comprises 24 closed questions related to the following dimensions of burden stemming from caregiving: time-dependence, developmental, social, emotional (five items each), and physical burden (four items). The inventory has a high Cronbach's alpha value of 0.936 and thus is a measure of internal consistency and reliability. Each item is scored from 0 (not at all) to 4 (highly descriptive); the higher the score the greater caregiver burden; there are no cutoff points for classifying burden. Therefore, total scores for dimensions 1–4 can range from zero to 20. For physical burden consisting of four items only, the equivalent score is reached by multiplying the summary result in this domain by 1.25 (Valer et al. 2015).

Specific members of the hospital physiotherapy team were appointed to maintain communication and counseling with the prime caregivers to the spinal cord injured patients. The same staff did the evaluation and counseling, related to the rehabilitative aspects, during the first few follow-ups. The CBD questionnaire was performed by the same staff 2 months after the patients had been discharged from the hospital. The caregivers were informed about the scope and aim of the study related to getting insight into the current health perspectives of chronically dependent neurological patients and establishing new strategies of management. The needful considerations such as the use of native language, high priority to caregivers' autonomy, adherence to ethical standards, and confidentiality of answers were strictly considered. Caregivers answered questionnaire in isolation to avoid subjecting them to any potential bias or influence.

Inclusion criteria:

- Primary caregivers
- Aged 18 and more
- At least 2 months of caregiving to spinal cord injured patients
- Unpaid care providers

Exclusion criteria:

- Professional experience in caregiving to chronically ill persons
- Interest in only financial and/or psychological patient support

Sample Size Calculation In Nepal, two lakhs (200,000) of disability cards were distributed by the government authorities during the fiscal year 2017–2018. Among them, 86,362 were categorized as concerning people with severe forms of disability, i.e., highly dependent on caregivers. Of those people, 31,090 (15.5%) suffered from extreme disability due mostly to spinal cord injuries. The sample size required for adequate statistical elaboration was calculated according to the formula $n = Z^2 \times p \times q/d2$, where $Z = 1.96$ at 95% confidence interval, $p = 15.5\%$ prevalence of extreme disability, $q = 1\text{-}p$, and $d = 10\%$ margin of error. The minimum sample size was calculated at 51 subjects, whereas the total number of patients included in the study was 71 subjects. Data were presented as frequency distribution (counts and percentages).

3 Results

There were 58 (81.7%) men and 13 (18.3%) women with extreme disability in the study. They were taken care of by 46 (64.8%) male and 25 (35.2%) female caregivers, which might be a result of the joint family system of caretaking present in 69.0% of the study patients. Concerning the travel distance for caregivers to attend patients in the medical center, 67.6% of them had to go more than 50 km. Most of the primary caregivers in the study reported to be significantly burdened owing to persistent and protracted nature of caring for disabled patients. The CBD score showed that 57.7% of caregivers perceived severe burden, 21.2% perceived moderate, and 16.9% perceived extreme burden. Only did 4.2% of caregivers perceive no burden related to caring tasks (Fig. 1).

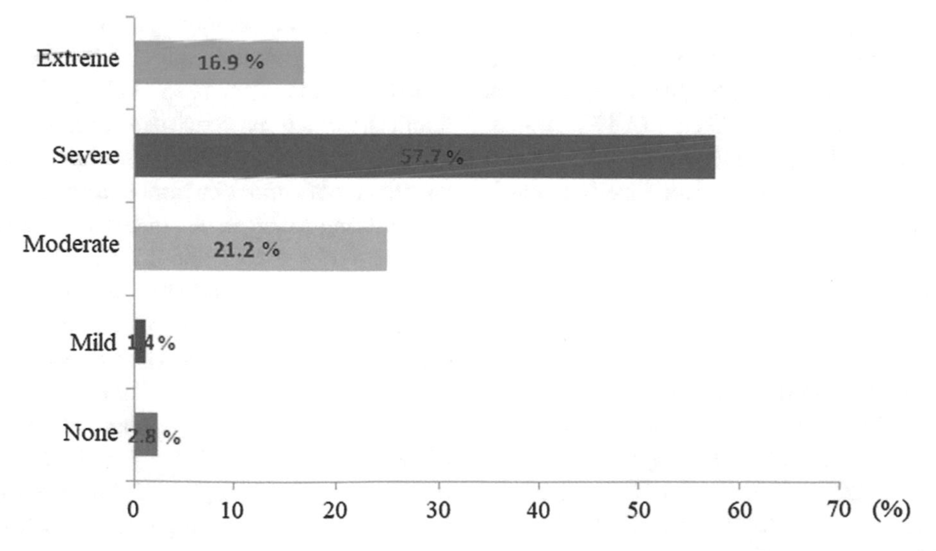

Fig. 1 Distribution of perceived caregivers' burden while caring for patients with extreme disability, assessed with Caregiver Burden Inventory (CBD)

Table 1 Tabulated results of significant domains pertaining to perception of caregivers' burden while caring for the severely disabled in Nepal

Variables	Parameters	Counts	%
Gender of patients	Male	58	81.7
	Female	13	18.3
Gender of primary caregivers	Male	46	64.8
	Female	25	35.2
Family patterns	Nuclear	22	31.0
	Joint	49	69.0
Distance from the care center (km)	< 50	24	33.8
	50–100	24	33.8
	> 100	23	32.4
Caretaker burden scale (score)	None	2	2.8
	Mild	1	1.4
	Moderate	15	21.1
	Severe	41	57.7
	Extreme	12	16.9

Additionally, a run-down analysis showed that financial matters had the greatest impact on the perception of caretaking burden, concerning 90.1% of caregivers. Moreover, 87.3% of caregivers had safety concerns, 67.6% worried about the future health of disabled relatives, a similar percentage felt physically strained or overtaxed with daily requirements of caring, and 56.0% expressed an opinion that that their social life was highly adversely affected by caretaking burden.

On the other hand, the variables bearing a minimum impact on the caregivers included concerns about their disturbed privacy (1.4%), health (2.8%), and the thoughts of leaving sick relatives to someone else's care (2.8%). A summary of the evaluation of the perceived burden of caretaking in the disabled individual's home is shown in Table 1.

4 Discussion

We found that most SCI patients were males (81.7%), which is in line with other similar studies (McDaid et al. 2019; Khazaeipour et al. 2017; Graça et al. 2013; Burns and O'Connell 2012). The male predominance is explicable by the risk-taking and more aggressive behavior of men, when compared with women, which makes them prone to accidents and injuries. Most primary caregivers, who attended chronically handicapped SCI patients, also were males (64.8%), which, however, contrasts the results of other studies where the caregivers were predominantly spouses of the male victims (Gopal et al. 2017; Lynch and Cahalan 2017; Gajraj-Singh 2011; Dreer et al. 2007; Unalan et al. 2001). The predominance of male caregivers in this study might stem from the cultural way of living in a joint family structure in Nepal, where taking care of a disabled individual becomes the responsibility of healthy male members in the family. Accordingly, our study shows that 69.0% of the caregivers belonged to joint families, which is consistent with a study of Bhattacharjee et al. (2012) who have found the prevalence of similar cultural and social beliefs.

Referring to the results of CBD inventory, the present study shows that 57.7% of caregivers perceived a severe burden of caretaking, 21.1% perceived moderate, and 16.9% perceived an extreme level of burden. Most previous studies have shown only mild-to-moderate burden of caring for the disabled (Gopal et al. 2017; Khazaeipour et al. 2017). The CBD scale is governed by the interplay of different variables such as the level of caregivers' education, age, duration of caregiving, severity of SCI handicap, and foremost their economic status. We found in this study that 90.1% of caregivers dealt with the issues relating to financial aspects of the situation, which is in line with the findings of other studies (Øderud 2014). The costs involved in the management of SCI, from the initial treatment to the prolonged rehabilitative phase, is high, even in the context of high-income countries (McDaid et al. 2019). In Nepal, where most people still live below the poverty line, the government allocated fiscal support for the SCI patient of $1000 seems a drop in the ocean of the mammoth expenses of the treatment and rehabilitation of such patients (Shrestha 2014). Furthermore, most caregivers are compelled to leave their permanent jobs to mold themselves to the caregiving requirements. This just adds up to the agony already faced by the caregivers. The lack of adequate fiscal support for SCI treatment has also been pointed out by Øderud (2014). In such circumstances, cutting off the treatment costs and waving taxes could substantially reduce the heightened burden among the caregivers.

This study shows that most caregivers had to travel significant distances to accompany the disabled relatives for routine visits to the medical center. Nepal is a mountainous country, and relevant health services are not readily available in most of the remote areas. About 32.4% of patients had to travel for more than 100 km, paying a handsome amount for the ground ambulance service, adding up to their financial and psychological burden. Additionally, 67.6% of caregivers felt physically strained, which is consistent with the findings of other studies (Bhattacharjee et al. 2012; Schulz et al. 2009). Many of them had to make an emergent transition to that of a care provider role with minimum facilitation of learning the high dependency care of SCI patients. Most of caregivers exhibited worries about the future of the disabled relatives. This study also revealed a major deterioration of family life and social life of cares (Golics et al. 2013; Chen and Boore 2009). A sudden unexpected role transitions put them in fear, uncertainty, and sometimes loss of identity. In the whole process, they experience anxiety and depression and may become undetected ill persons themselves. Caregivers are unable to participate in valued activities and interests and face loneliness as they become socially aloof (Sheija and Manigandan 2005; Weitzenkamp et al. 1997). They also are in constant fear of the impending death of the relative they care for. The fear of loneliness and abandonment and of further life after the death of a loved one has a profound emotional drain (Figueiredo et al. 2019; Salzer et al. 2016; Ebrahimzadeh et al. 2013).

5 Conclusions

Considering a significant frequency of spinal cord injuries, the burden of palliative care of such patients has emerged as one of the exigent public health issues. Spinal cord injury confers multispectral damaging effects not only to the patients but also to caregivers, affecting the family life in an unprecedented manner. These aspects of collateral concerns onto the caregivers have seldom been given much attention in Nepal. Low socioeconomic status superadds to a shortage of rehabilitation facilities, further enhancing the ripple effects to the already compromised quality of life of this subset of the population. The findings of our study demonstrate that most caregivers suffered from severe or extremely severe burden. In addition to the financial issues, there were numerous other interplaying factors adversely affecting the caregivers' lives. The meager fiscal help from the governmental side, paucity of rehabilitation centers, and discharging spinal cord-injured patients into the hands of untrained caregivers are the major concerns that merit immediate attention and remedial measures. It seems necessary to consider such patients and their caregivers as a unit and to provide the adequate counseling and support to both sides. The caregivers face extremely difficult time that may literally destroy their family, social, and work life, and in fact quite often they become undetected ill persons. It is time we made headways in reforming the issue marked by a sense of personal tragedy.

Conflicts of Interest The authors declare no conflicts of interest in relation to this article.

Ethical Approval All procedures performed in studies involving human participants were in accordance with the ethical standards of the national research committee and with the 1964 Helsinki declaration and its later amendments or comparable ethical standards. The study was approved by the Ethics Committee of the Nobel Medical College and Teaching Hospital in Biratnagar, Nepal (approval IRC-NMCTH 261/2019).

Informed Consent Written informed consent was obtained from all individual participants included in the study.

References

Bhattacharjee M, Vairale J, Gawali K, Dalal PM (2012) Factors affecting burden on caregivers of stroke survivors: population-based study in Mumbai (India). Ann Indian Acad Neurol 15:113–119

Burns AS, O'Connell C (2012) The challenge of spinal cord injury care in the developing world. J Spinal Cord Med 35:3–8

Chen HY, Boore JRP (2009) Living with a relative who has a spinal cord injury: a grounded theory approach. J Clin Nurs 18:174–182

Cripps RA, Lee BB, Wing P, Weerts E, Mackay J, Brown D (2011) A global map for traumatic spinal cord injury epidemiology: towards a living data repository for injury prevention. Spinal Cord 49:493–501

DATAVIZ (2016) Interactive map of the disabled population in Nepal. https://codefornepal.org/2016/07/disability/. Accessed on 7 May 2020

Ditunno JF, Young W, Donovan WH, Creasey G (1994) The international standards booklet for neurological and functional classification of spinal cord injury. Paraplegia 32:70–80

Dreer LE, Elliott TR, Shewchuk R, Berry JW, Rivera P (2007) Family caregivers of persons with spinal cord injury: predicting caregivers at risk for probable depression. Rehabil Psychol 52(3):351–357

Ebrahimzadeh MH, Shojaei BS, Golhasani-Keshtan F, Soltani-Moghaddas H, Fattahi AS, Mazloumi SM (2013) Quality of life and the related factors in spouses of veterans with chronic spinal cord injury. Health Qual Life Outcomes 11:48

Figueiredo R, Sá L, Lourenço T, Almeida S (2019) Death anxiety in palliative care: validation of the nursing diagnosis. Acta Paulista de Enfermagem 32:178–185

Gajraj-Singh P (2011) Psychological impact and the burden of caregiving for persons with spinal cord injury (SCI) living in the community in Fiji. Spinal Cord 49:928–934

Golics CJ, Basra MKA, Finlay AY, Salek S (2013) The impact of the disease on family members: a critical aspect of medical care. J R Soc Med 106:399–407

Gopal VV, Baburaj PT, Balakrishnan PK (2017) Caregiver's burden in rehabilitation of patients with neurological deficits following traumatic spinal cord injury. J Spinal Surg 4(1):9–13

Graça Á, do Nascimento MA, Lavado EL, Garanhani MR (2013) Quality of life of primary caregivers of spinal cord injury survivors. Rev Bras Enfermagem 66:79–84

Khazaeipour Z, Rezaei-Motlagh F, Ahmadipour E, Azarnia-Ghavam M, Mirzababaei A, Salimi N, Salehi-Nejad A (2017) Burden of care in primary caregivers of individuals with spinal cord injury in Iran: its association with sociodemographic factors. Spinal Cord 55:595–600

Lynch J, Cahalan R (2017) The impact of spinal cord injury on the quality of life of primary family caregivers: a literature review. Spinal Cord 55:964–978

Marvardi M, Mattioli P, Spazzafumo L, Mastriforti R, Rinaldi P, Polidori MC, Cherubini A, Quartesan R, Bartorelli L, Bonaiuto S, Cucinotta D, Di Iorio A, Gallucci M, Giordano M, Martorelli M, Masaraki G, Nieddu A, Pettenati C, Putzu P, Solfrizzi V, Tammaro AE, Tomassini PF, Vergani C, Senin U, Mecocci P, Study Group on Brain Aging; Italian Society of Gerontology and Geriatrics (2005) The Caregiver Burden Inventory in evaluating the burden of caregivers of elderly demented patients: results from a multicenter study. Aging Clin Exp Res 17(1):46–53

McDaid D, La Park A, Gall A, Purcell M, Bacon M (2019) Understanding and modeling the economic impact of spinal cord injuries in the United Kingdom. Spinal Cord 57:778–788

Middleton JW, Simpson GK, De Wolf A, Quirk R, Descallar J, Cameron ID (2014) Psychological distress, quality of life, and burden in caregivers during community reintegration after spinal cord injury. Arch Phys Med Rehabil 95:1312–1319

Øderud T (2014) Surviving spinal cord injury in low-income countries. Afr J Disabil 3(2):80

Salzer V, Stadtlander LM, Plebani J, Semenova V (2016) Death anxiety, depression, and coping in family caregivers. J Soc Behav Health Sci 10:34–48

Schulz R, Czaja SJ, Lustig A, Zdaniuk B, Martire LM, Perdomo D (2009) Improving the quality of life of caregivers of persons with spinal cord injury: a randomized controlled trial. Rehabil Psychol 54:1–15

Sheija A, Manigandan C (2005) Efficacy of support groups for spouses of patients with spinal cord injury and its impact on their quality of life. Int J Rehabil Res 28(4):379–383

Shrestha D (2014) Traumatic spinal cord injury in Nepal. Kathmandu Univ Med J 12:161–162

Unalan H, Gençosmanoğlu B, Akgün K, Karamehmetoğlu S, Tuna H, Ones K, Rahimpenah A, Uzun E, Tüzün F (2001) Quality of life of primary caregivers of spinal cord injury survivors living in the community: a controlled study with short form-36 questionnaire. Spinal Cord 39(6):318–322

Valer DB, Aires M, Fengler FL, Paskulin LM (2015) Adaptation and validation of the Caregiver Burden Inventory for use with caregivers of elderly individuals. Rev Lat Am Enfermagem 23(1):130–138

Weitzenkamp DA, Gerhart KA, Charlifue SW, Whiteneck GG, Savic G (1997) Spouses of spinal cord injury survivors: the added impact of caregiving. Arch Phys Med Rehabil 78:822–827

World Bank (2019). https://datahelpdesk.worldbank.org/knowledgebase/articles/906519-world-bank-country-and-lending-groups. Accessed on 3 May 2020

Printed in the United States
by Baker & Taylor Publisher Services